Computational Analysis
with the
HP 25 Pocket Calculator

Computational Analysis with the HP 25 Pocket Calculator

PETER HENRICI *1923-*

Professor of Mathematics
Eidgenössische Technische Hochschule, Zürich

A Wiley-Interscience Publication
JOHN WILEY & SONS
New York · London · Sydney · Toronto

Library of Congress Cataloging in Publication Data:
Henrici, Peter, 1923-
  Computational analysis with the HP 25 pocket calculator.

  "A Wiley-Interscience publication."
  1. Numerical analysis—Data processing. 2. Numbers,
Theory of—Data processing. 3. Calculating-machines.
I. Title. II. Title: HP 25 pocket calculator.

QA297.H399   519.4   77-1182
ISBN 0-471-02938-6

Printed in the United States of America

10 9 8 7 6 5 4 3

PREFACE

This brochure contains some 30 programs written for a specific programmable pocket calculator, the HP-25. These programs implement algorithms in number theory, equation solving, algebraic stability theory, numerical integration, as well as for the evaluation of special functions such as the gamma function, various Bessel functions, and the Riemann zeta function. By means of the flow diagrams and the detailed descriptions that are provided, the programs are easily adapted to run on any calculator of comparable capacity.

The purpose to be served by these programs is didactic as well as practical.

The immediate didactic purpose is to enable students of numerical and computational analysis to gain first-hand experience with some modern techniques of scientific computing. I have used a number of the programs in the classroom to demonstrate the numerical performance of such techniques, using data provided spontaneously by the students. Such demonstrations greatly increase the practical, "know-how" content of numerical analysis courses. They also eliminate the suspicion that the instructor uses examples that are rigged to show an algorithm in an especially favorable light.

My purpose is didactic in another, potentially even more important sense. Programmable calculators have now been in use for almost a third of a century. Yet many members of the scientific community, although they use mathematics in their work, have remained ignorant of the essentials of modern scientific computation. This even holds for some mathematicians, if they did not choose to specialize in numerical analysis or computer science. All this is

destined to change with the advent of the programmable pocket calculator. By providing the basic fea-
tures of branching and iteration, these amazing in-
struments put automatic computation within the reach,
in the privacy of one's own study, of anybody who re-
members his basic calculus. No special programming
language must be learned to operate one of these de-
vices; to its owner, the access is immediate, and
there are no waiting times beyond those required by
the computation itself. An error in the program or in
the data can be corrected immediately. Thus these
calculators are interactive to a degree that up to
now was enjoyed only by the privileged few who were
connected to an expensive computer by their own ter-
minal. By publishing my programs I hope to bring some
of the flavor of modern automatic computation to
those who so far have been untouched by it.

Finally, some of my programs, especially those
dealing with special functions, also serve an emi-
nently practical purpose. If I quickly need some values of a higher transcendental function such as a
Bessel function of nonintegral order, I know of no
way to get them with less fuss than by using my
little calculator. Similar programs could (and un-
doubtedly will) be written for functions other than
those represented in my collection; my programs merely provide a small sample of what can be done in this
area.

Coming back to the didactic aspect, it must be
admitted that not all areas of modern numerical computation can be demonstrated satisfactorily on a
pocket calculator. Because of the limited memory capacity, problems dealing with large sets of data,
such as computations in linear algebra, optimization,
or partial differential equations, cannot be solved
on a pocket calculator as they would be solved on a
large computer. Even in problems requiring small sets
of data, the lack of memory space imposes limitations.
Thus it happens, for instance, that my programs for
determining the zeros or the stability properties of
polynomials in most cases can deal only with polynomials of degree not exceeding four. Such programs ne-
vertheless can serve as models - or "pilot programs"
- for similar programs designed to deal with larger

sets of data on correspondingly larger computers. Indeed, some of my programs for the evaluation of special functions would be written in much the same way also for the very largest computers; the only difference here lies in the speed of execution.

Naturally, this collection is not meant to be exhaustive. The creative student can find ample room for the design of new programs, particularly in the areas of equation solving, numerical integration, and the evaluation of special functions. Furthermore, it is likely that many of the programs presented here could be improved in some way, say, by making them faster, easier to use, or more automatic. I will be genuinely pleased to learn about such improvements.

Already when compiling these programs I have benefitted from the advice and help of numerous collegues, students, and friends. J. Waldvogel and D. D. Warner contributed ideas. W. Seewald and A. Stähli suggested improvements in existing programs. E. Specker offered a solution to the problem of writing viable power series programs. My wife, Marie-Louise Henrici, did most of the checking and proofreading and eliminated numerous errors. The reader may judge for himself the excellent work of Brigitte Knecht who transformed my manuscript into camera-ready copy. The staff of John Wiley & Sons disposed of all problems of editing and production with professional know-how. To all these individuals, and to any other helpers who may not have been named here, I offer my heartfelt thanks.

Peter Henrici

Zurich, Switzerland
February 1977

CONTENTS

x   Contents

Computational Analysis
with the
HP 25 Pocket Calculator

INTRODUCTION

This is not a textbook on the programming of pocket calculators. To learn how to operate your calculator, consult your manual. As a first introduction to the manifold uses of the pocket calculator, the recent book by Jon M. Smith, "Scientific analysis on the pocket calculator" (Wiley, 1975) can be recommended.

Each program description is arranged according to the following scheme:

1. Purpose,
2. Method,
3. Flow diagram,
4. Storage and program,
5. Operating instructions,
6. Examples and timing.

The statement of purpose usually is very brief. It is intended to state, in language as nontechnical as possible, what the program is good for.

The description of the method is strictly confined to what is being done. For the analytical justification and for technical details, the reader is referred to the literature. References to books by the author are given in the following abbreviated form:

ENA   = "Elements of numerical analysis",
             Wiley, New York, 1964,

ACCA I = "Applied and computational complex ana-
             lysis", Vol. I, Wiley, New York, 1974,

ACCA II = "Applied and computational complex ana-
             lysis", Vol. II, Wiley, New York, 1977.

The flow diagrams do not follow any particular
standard format. The intent is simply to make the
structure of a program visible at a glance. In some
cases explanations are given for the technical details of a program or for abbreviations that are
used in the flow diagram.

The section entitled storage and program, naturally, is the core part of each program description. We
use the symbol R_k to denote the storage register numbered k. Quantities that must be stored in the storage
registers ahead of the computation are encased, thus:

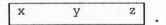 .

The program listing uses the key symbols as they are
found on the keys of your HP-25 calculator, with the
exception of multiplication, which is indicated by *
in order to avoid confusion with the letter "x". The
listing is as compact as possible. For yellow and
blue instructions it is understood that the keys "f"
and "g" are to be pressed in advance; no ambiguity
can arise by our abbreviated notation. The arrows →

in front of certain instruction numbers are not part
of the program. They simply indicate an entrance from
a GTO instruction. For those readers who would have
liked to see more complete descriptions of the technical details of a program, it must be said that such
paraphrases can make for extremely tedious reading.
The best way to understand the details of a program
is for the user to try to write his own. He then will
often realize why the particular limitations of the
pocket calculator forced the author to deviate from
the more straightforward ways in which such programs
would be written for a larger computer.

An attempt has been made to keep the operating
instructions self-contained, except in a few obvious
cases. Here we also give indications and explanations
of possible failures of some programs.

The examples are not intended to be particularly
sophisticated. The user should try one of the simpler
examples to see whether he has correctly loaded the
program and followed the instructions. In some cases
the examples indicate how the program would be used
in the classroom, illustrating particular points of
analysis. All timings given are approximate.

Part 1
NUMBER THEORY

PRIME FACTOR DECOMPOSITION

1. Purpose

To produce, for any positive integer $m < 10^9$, its decomposition in prime factors. Factors are to appear in
increasing order.

2. Method

We divide m successively by the numbers

$$d = 2, 3, 5, 7, 11, 13, 17, 19, 23, 25, 29, 31, \ldots,$$

that is, by 2, 3 and by all odd integers > 3 that are
not divisible by 3, until either (a) $d > \sqrt{m}$ or (b) the
remainder is zero. In case (a), m is prime; the calculator shows m and then zero to indicate that the decomposition is terminated. In case (b), d is a divisor
of m. The calculator shows d, replaces m by m/d, and
starts again, using the last d as its first divisor.

To compute the d, we start with d = 2 and calculate the successor d' of d by the formula

$$
d' = \begin{cases} 2d - 1 \text{ , if } d \leqq 4 \\[2em] d + 3 + \varepsilon \text{ , if } d > 4 \end{cases}
$$

Here the value of ε alternates between +1 and -1, beginning with -1.

3. <u>Flow Diagram</u>

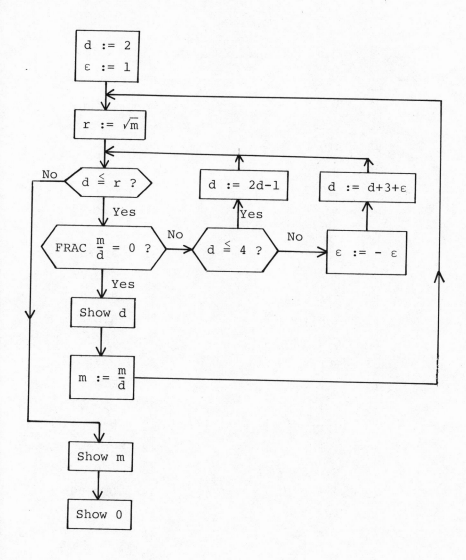

4. Storage and Program

R_0	R_1	R_2	R_3	R_4	R_5	R_6	R_7
m	$\sqrt{m}$	d	ε				

	00			25	1
	01	STO 0		26	STO-2
	02	2		27	GTO 09
	03	STO 2	→	28	3
	04	1		29	STO+2
	05	STO 3		30	RCL 3
	06	RCL 0		31	CHS
→	07	$\sqrt{x}$		32	STO 3
	08	STO 1		33	STO+2
→	09	RCL 2		34	GTO 09
	10	RCL 1	→	35	RCL 0
	11	x < y		36	RCL 2
	12	GTO 41		37	R/S
	13	RCL 0		38	÷
	14	RCL 2		39	STO 0
	15	÷		40	GTO 07
	16	FRAC	→	41	RCL 0
	17	x = 0		42	R/S
	18	GTO 35		43	0
	19	4		44	GTO 00
	20	RCL 2		45	
	21	x $\overset{\geq}{=}$ y		46	
	22	GTO 28		47	
	23	2		48	
	24	STO*2		49	

5. Operating Instructions

After loading the program, move the operating switch to RUN. Then press

FIX 0

to get a nine-digit display of all integers. If prime factors of m are desired, load m into the X register. When

PRGM
R/S

is pressed, the calculation will start, then stop by displaying smallest prime factor of m. When

R/S

is pressed, the next prime factor will be displayed, and so forth, until 0 is shown, indicating that all prime factors have been found.

6. Examples and Timing

[1]   36 = 2 * 2 * 3 * 3. Time required about 8 sec.
[2]   71489 = 11 * 67 * 97. Time required about 28 sec.
[3]   987654321 = 3 * 3 * 17 * 17 * 379721. Time required about 220 sec.

EUCLIDEAN ALGORITHM

1. Purpose

To determine the greatest common divisor of two integers a and b, not both zero, $|a|$, $|b| < 10^9$.

2. Method

The Euclidean algorithm. We determine a sequence of nonnegative integers $\{n_i\}$ by

$$n_0 := \max (|a|,|b|) \, ,$$

$$n_1 := \min (|a|,|b|) \, ,$$

$$n_k := n_{k-2} - \left[\frac{n_{k-2}}{n_{k-1}} \right] n_{k-1} \, ,$$

k = 2, 3, ... This sequence is decreasing, and thus it eventually reaches zero. If $n_k = 0$, then

$$c := n_{k-1}$$

is the greatest common divisor of a and b (see ACCA II, § 12.2).

3. Flow Diagram

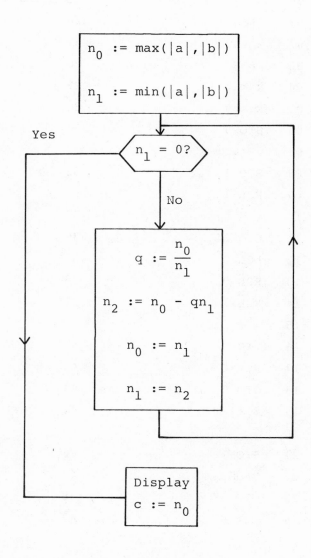

4. Storage and Program

R_0	R_1	R_2	R_3	R_4	R_5	R_6	R_7
a	b						
n_0	n_1						

00		25	GTO 00
01	ABS	26	
02	$x \gtrless y$	27	
03	ABS	28	
04	$x < y$	29	
05	$x \gtrless y$	30	
06	STO 0	31	
07	R ↓	32	
08	STO 1	33	
→ 09	RCL 1	34	
10	$x = 0$	35	
11	GTO 24	36	
12	RCL 0	37	
13	RCL 0	38	
14	RCL 1	39	
15	÷	40	
16	INT	41	
17	RCL 1	42	
18	STO 0	43	
19	*	44	
20	-	45	
21	STO 1	46	
22	PAUSE	47	
23	GTO 09	48	
→ 24	RCL 0	49	

5. Operating Instructions

Load the program. Switch to RUN. Press

FIX 0

to get integers displayed as integers. Load a into R_0 and b into R_1. Press

PRGM

R/S

to start computation. The calculator will stop by displaying c, the greatest common divisor of a and b. If both a and b are zero, c = 0.

6. Examples and Timing

$\boxed{1}$ a = 45, b = 96 yields c = 3. Computing time about 3 sec.

$\boxed{2}$ a = - 965302379, b = 980051 yields c = 997, the correct result. Computing time about 5 sec.

$\boxed{3}$ a = 1, b = 0 yields c = 1.

RATIONAL BINOMIAL COEFFICIENTS

1. Purpose

To represent, for arbitrary rational $\rho = \frac{p}{q}$, the binomial coefficients

$$b_0(\rho) := \binom{\rho}{0} := 1 \; ,$$

$$b_n(\rho) := \binom{\rho}{n} := \frac{\rho(\rho-1)(\rho-2)\ldots(\rho-n+1)}{n!}$$

$n = 1, 2, \ldots$, as irreducible rational fractions,

$$b_n(\rho) = \frac{r_n}{s_n} \; , \tag{1}$$

where r_n and s_n are integers, $(r_n, s_n) = 1$, and $s_n > 0$. Such rational values of the binomial coefficients are often needed in analytical work involving the binomial series,

$$(1+x)^\rho = \sum_{n=0}^{\infty} b_n(\rho)\, x^n \; , \quad -1 < x < 1 \; .$$

2. Method

The recurrence relation $b_0(\rho) = 1$,

$$b_{n+1}(\rho) = \frac{\rho-n}{n+1} b_n(\rho) \ , \ n = 0, 1, 2, \ldots \ ,$$

is used. If ρ is a rational number, $\rho = \frac{p}{q}$, and if b_n is represented in the form (1), this yields

$$\frac{r_{n+1}}{s_{n+1}} = \frac{p - qn}{q(n+1)} \frac{r_n}{s_n} \ .$$

We thus first compute the integers

$$r^*_{n+1} := (p - qn)r_n \ , \ s^*_{n+1} := q(n+1)s_n \ ,$$

then by means of the Euclidean algorithm determine their greatest common divisor

$$c := (r^*_{n+1}, \ s^*_{n+1})$$

and find

$$r_{n+1} = \frac{r^*_{n+1}}{|c|} \ , \ s_{n+1} = \frac{s^*_{n+1}}{|c|} \ .$$

The process is started with $r_0 = s_0 = 1$. The short version of the Euclidean algorithm used here is borrowed from the program "Exact continued fractions for quadratic irrationalities".

3. <u>Flow Diagram</u>

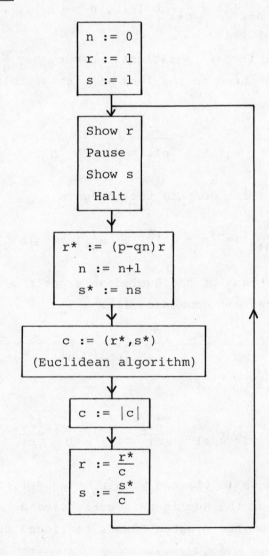

4. Storage and Program

R_0	R_1	R_2	R_3	R_4	R_5	R_6	R_7
p	q	r,r*	s,s*	n		r*	s*

	00		25	STO 7
	01	CLX	→ 26	RCL 6
	02	STO 4	27	RCL 6
	03	1	28	RCL 7
	04	STO 2	29	÷
	05	STO 3	30	INT
→	06	RCL 2	31	RCL 7
	07	PAUSE	32	STO 6
	08	RCL 3	33	*
	09	R/S	34	-
	10	RCL 0	35	STO 7
	11	RCL 1	36	x ≠ 0
	12	RCL 4	37	GTO 26
	13	*	38	RCL 6
	14	-	39	ABS
	15	STO*2	40	STO÷2
	16	1	41	STO÷3
	17	STO+4	42	GTO 06
	18	RCL 1	43	
	19	RCL 4	44	
	20	*	45	
	21	STO*3	46	
	22	RCL 2	47	
	23	STO 6	48	
	24	RCL 3	49	

5. Operating Instructions

Load the program; move operating switch to RUN. Press

FIX

0

to obtain integer display. Load numerator p of ρ into
R_0 and denominator q into R_1. On pressing

PRGM

R/S

calculator briefly displays $r_0 = 1$ and halts by displaying $s_0 = 1$. Pressing

R/S

will cause display of r_1 and, after a short pause,
s_1, etc. Pressing

÷

after display of s_n will cause the decimal value of

$$b_n(\rho) = \frac{r_n}{s_n}$$

to be displayed, without disturbing subsequent computations. Also, after display of s_n, the current value of n may be exhibited by pressing

$$RCL\ 4$$

To re-inspect r_n after display of s_n, press

$$x \lessgtr y$$

For large n, integer overflow may cause the values of r_n and s_n to be inaccurate; however, the decimal values of $b_n(\rho)$ will still be accurate.

6. Examples and timing

⒈   For $\rho = 6$, i.e. p = 6, q = 1, we get
  $\binom{6}{n} = 1,\ 6,\ 15,\ 20,\ 15,\ 6,\ 1,\ 0,\ 0,\ 0,\ \ldots$.

⒉ $\rho = \frac{1}{2}$, i.e. p = 1, q = 2 yields the values

$$\binom{1/2}{n} = 1,\ \frac{1}{2},\ -\frac{1}{8},\ \frac{1}{16},\ -\frac{5}{128},\ \frac{7}{256},\ -\frac{21}{1024},$$

$$\frac{33}{2048},\ -\frac{429}{32768},\ \frac{715}{65536},\ -\frac{2431}{262144},\ \frac{4199}{524288}$$

$$-\frac{29393}{4194304},\ \frac{52003}{8388608},\ -\frac{185725}{33554432},$$

$$\frac{334305}{67108864},\ -\frac{9694845}{2147483648},\ \frac{20036013}{4867629602}.$$

Computing time approx. 165 sec.

CONTINUED FRACTION REPRESENTATIONS OF REAL NUMBERS

1. Purpose

Given a real number ρ, to find its standard representation either by a terminating continued fraction,

$$\rho = b_0 + \cfrac{1}{b_1} + \cfrac{1}{b_2} + \ldots + \cfrac{1}{b_n} , \qquad (1)$$

if ρ is rational, or by a nonterminating continued
fraction,

$$\rho = b_0 + \cfrac{1}{b_1} + \cfrac{1}{b_2} + \ldots , \qquad (2)$$

if ρ is irrational. Here b_0 is an integer, and the
b_i ($i = 1, 2, \ldots$), if defined, are positive integers.

2. Method

By denoting by $[x]$ the greatest integer not exceeding
x, the b_i may be computed as follows (see ACCA II,
§ 12.2): Let

$$b_0 := [\rho] \ , \ \rho_1 := \rho - b_0 \ ,$$

and for $k = 1, 2, \ldots$, if $\rho_k \neq 0$,

$$b_k := \left[\frac{1}{\rho_k}\right] \ , \ \rho_{k+1} := \frac{1}{\rho_k} - b_k \ . \tag{3}$$

If ρ is an integer, then already $\rho_1 = 0$, and the representation (1) reduces to $\rho = b_0$. If ρ is rational but not an integer, then eventually $\rho_{n+1} = 0$ for some $n > 0$. The $b_0, b_1, \ldots, b_n$ thus generated are the partial denominators of the continued fraction (1). If ρ is irrational, then $\rho_k \neq 0$ for all $k \geq 1$.

3. Flow Diagram

A slight complication arises through the fact that the operation INTρ agrees with $[\rho]$ only if $\rho \geq 0$, or if $\rho < 0$ and ρ is an integer. For nonintegral $\rho < 0$ we have

$$[\rho] = \text{INT}\rho - 1 \ .$$

Otherwise, the flow diagram is straightforward.

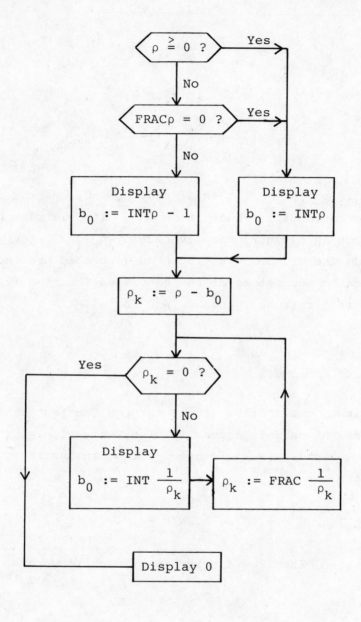

4. Storage and Program

R_0	R_1	R_2	R_3	R_4	R_5	R_6	R_7

ρ

00			25	FRAC
01	STO 0		26	GTO 18
02	$x \gtreqless 0$	→	27	R/S
03	GTO 12		28	GTO 27
04	FRAC		29	
05	x = 0		30	
06	GTO 12		31	
07	RCL 0		32	
08	INT		33	
09	1		34	
10	-		35	
11	GTO 14		36	
→ 12	RCL 0		37	
13	INT		38	
→ 14	R/S		39	
15	RCL 0		40	
16	$x \gtrless y$		41	
17	-		42	
→ 18	x = 0		43	
19	GTO 27		44	
20	1/x		45	
21	ENTER		46	
22	INT		47	
23	R/S		48	
24	R ↓		49	

5. Operating Instructions

Load the program; switch to RUN. Load ρ into the X register. Press

FIX 0

to get the integer display. By pressing

PRGM

R/S

b_0 will be displayed. Pressing

R/S

repeatedly will display b_1, b_2, ... If a zero is displayed, this means that the algorithm has terminated, the number ρ being rational. Note, however, that because of rounding errors the algorithm does not always operate as predicted by the theory. Thus for rational ρ we may sometimes get an exceedingly large b_{n+1} in place of 0, indicating that on the computer ρ_{n+1} was not exactly zero as it should have been. For irrational ρ, the b_i from a certain point onward generally are wrong. These deficiencies could only be corrected by working with multiple precision (complicated on the HP 25) or, in certain special cases, with integer arithmetic (see the following program).

6. Examples

[1] $\rho = 2/3$ yields $b_k = 0, 1, 2, 0$. Indeed,

$$\frac{2}{3} = \frac{1}{\lfloor 1} + \frac{1}{\lfloor 2} .$$

[2] $\rho = -4.5$ yields $b_k = -5, 2$. Indeed,

$$-4.5 = -5 + \frac{1}{2} .$$

[3] $\rho = 26/47$ yields $b_i = 0, 1, 1, 4, 4, 1,$
(20000000). The last entry is wrong. In fact,

$$\frac{26}{47} = \frac{1}{\lfloor 1} + \frac{1}{\lfloor 1} + \frac{1}{\lfloor 4} + \frac{1}{\lfloor 4} + \frac{1}{\lfloor 1} .$$

[4]   For $\rho = e$ we get, as discovered by Euler, $b_i =$

2, 1, 2, 1, 1, 4, 1, 1, 6, 1, 1, 8, 1, 1, (3).

The last entry should be 10 and hence is wrong.

[5] $\rho = \pi$ furnishes $b_i =$

3, 7, 15, 1, (293) .

The first few approximants of the continued fraction for π thus are

$$3 + \frac{1}{7} \qquad\qquad = \frac{22}{7} = 3.14285714 \;,$$

$$3 + \frac{1}{\lfloor 7} + \frac{1}{\lfloor 15} \qquad = \frac{333}{106} = 3.14150943 \;,$$

$$3 + \frac{1}{\lfloor 7} + \frac{1}{\lfloor 15} + \frac{1}{\lfloor 1} = \frac{355}{113} = 3.14159292 \;.$$

The next approximant, unfortunately, is already wrong (b_4 should be 292), which is indicative of the great accuracy that is already achieved by the foregoing low order approximants.

6 $\rho = \dfrac{1 + \sqrt{5}}{2}$. We get $b_0 = b_1 = \ldots = b_{21} = 1$, and then $b_{22} = 2$, which is wrong.

7 $\rho = \sqrt{7}$. The sequence of b_i's is as follows:

2, 1, 1, 1, 4, 1, 1, 1, 4, 1, 1, 1, 4, 1, 1, 1, (3)

The last two examples illustrate the fact that the continued fractions representing quadratic irrationalities are ultimately periodic. The following program shows how to construct them without rounding error.

EXACT CONTINUED FRACTIONS FOR
QUADRATIC IRRATIONALITIES

1. Purpose

To construct, without rounding error, the standard
continued fraction representation of a quadratic irrationality, that is, of a real number

$$\rho = \frac{p + q\sqrt{r}}{m} ,$$

(1)

where p, q, m, r are integers and r is not a square,
m ≠ 0. The set of quadratic irrationalities coincides
with the set of irrational solutions of quadratic
equations $ax^2 + bx + c = 0$ with integer coefficients
a, b, c. By a celebrated theorem of Lagrange (see
ACCA II, § 12.2) the standard continued fraction representation of a quadratic irrationality is ultimately periodic. It thus suffices to determine, in addition to the nonperiodic part, one period of the continued fraction.

2. Method

The same division algorithm as in the preceding algorithm, but performed in rational arithmetic. If r is
fixed, the numbers of the form (1) create a field F;
that is, sums, differences, products, and quotients
of such numbers can again be represented in the same
way. This is clear for sums, differences, and products; for quotients, we may use the identity

$$\frac{1}{\rho} = \frac{m}{p + q\sqrt{r}} = \frac{-mp + mq\sqrt{r}}{q^2 r - p^2} \ .$$

Each number $\rho \ \varepsilon$ F thus is represented by a triple of
integers $[p, q, m]$. Many triples represent the same ρ,
but among all triples representing ρ there is a unique
representation, which we call the reduced representation, that is characterized by the fact that p, q, m
have no common factors and that m > 0. We may use the
symbol

$$[p, \ q, \ m] \ \backsim \ \rho$$

for any representation of ρ, and

$$[p, \ q, \ m] \ \tilde{\backsim} \ \rho$$

for the reduced representation. For any extended numerical work in F one should use reduced representa-

tions, both to prevent overflow and to recognize repeated elements. To this end, the integers p, q, m
should be divided after each operation in F by their
greatest common divisor (g.c.d.), say, d. This is determined by first calculating the g.c.d. (p,m) of p
and m by the Euclidean algorithm, and then construc-
ting d = ((p,m),q) by another application of the Euclidean algorithm.

For our present purposes the division algorithm
is formulated as follows: Let α_0 := ρ, and for k = 0,
1, 2, ... compute

$$b_k := [\alpha_k] \, ,$$

$$\alpha_{k+1} := \frac{1}{\alpha_k - b_k} \, .$$

There is no possibility of vanishing denominators, because the continued fraction is a priori known to be
infinite. The period of the continued fraction is com-
pleted as soon as $\alpha_{k+n} = \alpha_k \geq 0$ and n > 0, that is,
as a reduced triple representing α_k is repeated.

3. Flow Diagram

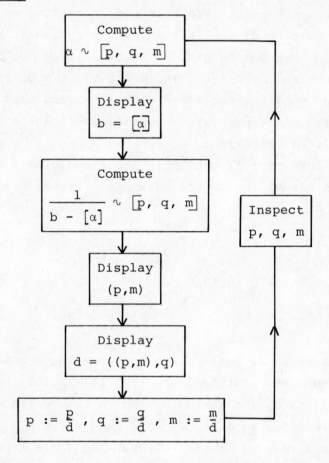

To save programming space, the program computes b_0 correctly only if $\rho > 0$. For similar reasons, a branching after the second application of the Euclidean algorithm must be performed manually.

4. Storage and Program

R_0	R_1	R_2	R_3	R_4	R_5	R_6	R_7
p	q	m	r	temp		temp	temp

00			25	R ↓
01	RCL 3		26	STO 2
02	√x		27	STO 7
03	RCL 1		28	RCL 0
04	*		29	STO 6
05	RCL 0	→	30	RCL 6
06	+		31	RCL 6
07	RCL 2		32	RCL 7
08	÷		33	÷
09	INT		34	INT
10	R/S		35	RCL 7
11	RCL 2		36	STO 6
12	*		37	*
13	STO-0		38	-
14	RCL 1		39	STO 7
15	x^2		40	x ≠ 0
16	RCL 3		41	GTO 30
17	*		42	RCL 6
18	RCL 0		43	R/S
19	x^2		44	RCL 1
20	-		45	STO 7
21	RCL 2		46	GTO 30
22	STO*1		47	STO÷0
23	CHS		48	STO÷1
24	STO*0		49	STO÷2

A short version of the Euclidean algorithm is con-
tained in instructions 30 through 43.

5. Operating Instructions

Load the program, then move the operating switch to
RUN. Press

$$FIX \ 0$$

to get the integer display. Load

$$p \ \text{into} \ R_0$$
$$q \ \text{into} \ R_1$$
$$m \ \text{into} \ R_2$$
$$r \ \text{into} \ R_3$$

The following instructions should now be carried out
cyclically. *Press

$$PRGM$$
$$R/S$$

The computer very shortly will display b_0. Pressing

$$R/S$$

will cause program to compute nonreduced representation $[p, \ q, \ m]$ of $1/(\alpha_0 - b_0)$ and to compute and dis-

play (p,m). Pressing

$$R/S$$

once more causes display of d = ((p,m),q). Now press

$$GTO\ 47$$
$$R/S$$

This will compute the reduced representation $\boxed{p,q,m}$
$\approx \alpha_1$. The calculator will stop by displaying d. To
inspect reduced representation, press

$$RCL\ 0$$
$$RCL\ 1$$
$$RCL\ 2$$

This will in turn show p, q, m. Now go back to * to
compute b_1 and so forth. The period of the fraction
is complete if a reduced triple p,q,m is repeated;
the period begins where the repeated triple occurs
the first time.

6. Examples and Timing

$\boxed{1}$ $\rho = \dfrac{24 - \sqrt{15}}{17}$. We obtain

b_i	1	5	2	3	2
p_i	24	7	3	3	3
q_i	-1	1	1	1	1
m_i	17	2	3	2	3

The repeated triple is encased. Indicating the period by a bar, the sequence of the b_i thus is

$$\{ 1, 5, \overline{2, 3} \} .$$

2 $\rho = \dfrac{4 + \sqrt{5}}{7}$. Here the results are

b_i	0	1	8	6	7	1	1
p_i	4	28	17	15	15	13	6
q_i	1	-7	7	7	7	7	7
m_i	7	11	4	5	4	19	11

b_i	1	30	1	1	1	7	6
p_i	5	15	15	5	6	13	15
q_i	7	7	7	7	7	7	7
m_i	20	1	20	11	19	4	5

Total computing time about 10 min.

Part 2
ITERATION

ITERATION

1. Purpose

To determine a fixed point of a given function f,
that is, a solution of the equation

$$x = f(x). \tag{1}$$

This may also be used to construct solutions of equations of the form

$$g(x) = 0 \tag{2}$$

by seeking fixed points of the function

$$f(x) := x + c \, g(x),$$

where c is a suitably chosen constant or function.

2. Method

We construct an iteration sequence $\{x_n\}$ by choosing
x_0 arbitrarily and forming

$$x_{n+1} = f(x_n) \ , \ n = 0, \ 1, \ 2, \ \ldots \ . \qquad (3)$$

If f maps a closed interval I into itself, if f is
contracting on I (i.e., if there exists k < 1 such
that $|f(x) - f(y)| \leq k|x - y|$ for arbitrary x, y ε I),
it can be shown that the equation (1) has precisely
one solution s in I, and any iteration sequence started with an x_0 ε I converges to s (see ENA, § 4.2).

3. <u>Flow Diagram</u>

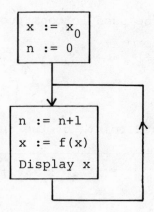

The index n merely counts the iteration steps; it is
not actually needed in the computation.

4. Storage and Program

R_0	R_1	R_2	R_3	R_4	R_5	R_6	R_7
x							n

00		08	PAUSE (R/S)
01	STO 0	09	GTO 04
02	CLX	10	
03	STO 7	11	
04	1	12	
05	STO+7	13	
06	GTO 10	14	
07	STO 0	15	

The program to compute f should be in the locations
10 through 49. The last instruction of this program
should be GTO 07.

5. Operating Instructions

Load the program; turn the switch to RUN. Select the
mode of displaying numbers, for example, by pressing

FIX 8

Load the starting value x_0 into the X register. Pressing

PRGM

R/S

will start the computation. The computer will pause
briefly after computation of each iterate and display
it. If a stop is desired, instruction 08 should be
R/S; in this case,

R/S

must be pressed after each display. The iteration
index n is displayed by pressing

RCL 7

after display of x. No convergence test is provided
by this simple program.

6. Examples and Timing

1   $f(x) := \sqrt{2 + x}$. The program to compute f is

10	RCL 0
11	2
12	+
13	$\sqrt{x}$
14	GTO 07

Starting with $x_0 := 0$, we get

$$x_1 = 1.41421356$$
$$1.84775907$$
$$1.96157056$$
$$1.99036945$$
$$1.99759091$$

. . .

$$x_{13} = 1.99999996$$
$$x_{14} = 1.99999999$$
$$x_{15} = 2.00000000$$

which, to the number of digits shown, equals s = 2. Computing time about 20 sec.

2 $f(x) := \dfrac{1}{1 + x}$. Again starting with $x_0 := 0$ we get

$$x_1 = 1.00000000$$
$$0.50000000$$
$$0.66666667$$
$$0.60000000$$

. . .

$$x_{19} = 0.61803400$$
$$x_{20} = 0.61803398$$
$$x_{21} = 0.61803399$$
$$x_{22} = 0.61803399$$

This equals s = $\frac{1}{2}(\sqrt{5} - 1)$, the unique positive solution of

$$x = \frac{1}{1 + x} \; .$$

Computing time about 30 sec.

3   $f(x) := (x - 1)^2$. With $x_0 := 3$ we get a rapidly divergent sequence; with $x_0 := 2$ we find

$$x_1 = 1.0000$$
$$x_2 = 0.0000$$
$$x_3 = 1.0000$$
$$x_4 = 0.0000 \; ,$$

an oscillating sequence.

4   $f(x) := e^{-10x}$. $x_0 := 0$ produces a sequence that, although theoretically convergent, does not converge. After n = 833 iterations, the sequence cycles between

$$x_{833} = 0.99954403 \quad \text{and}$$
$$x_{834} = 0.00004561 \; .$$

ITERATION WITH AITKEN ACCELERATION

1. <u>Purpose</u>

To determine a fixed point of a given function f more rapidly than with ordinary iteration.

2. <u>Method</u>

Aitken acceleration. Along with the sequence $\{x_n\}$, which is generated as in the preceding program, we generate the sequence of accelerated values $\{x_n'\}$, where

$$x_n' := x_n - \frac{(\Delta x_n)^2}{\Delta^2 x_n}$$

(Δ = forward difference operator). If $\{x_n\}$ converges, then under certain conditions (see ENA, § 4.4) $\{x_n'\}$ converges to the fixed point more rapidly than $\{x_n\}$.

3. <u>Flow Diagram</u>

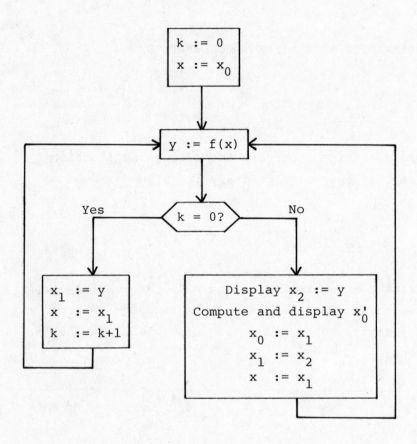

4. Storage and Program

R_0	R_1	R_2	R_3	R_4	R_5	R_6	R_7
$\boxed{x_0}$	x_1	x_0	Δx_0	$\Delta^2 x_0$	x	$f(x)$	k

00			25	STO 4		
01	CLX		26	RCL 3		
02	STO 7		27	x^2		
03	RCL 0		28	RCL 4		
04	STO 5		29	$\div$		
05	GTO 39		30	CHS		
→ 06	RCL 7		31	RCL 0		
07	$x \neq 0$		32	+		
08	GTO 15		33	R/S		
09	RCL 6		34	RCL 1		
10	STO 1		35	STO 0		
11	STO 5		36	RCL 2		
12	1		37	STO 1		
13	STO+7		38	STO 5		
14	GTO 39		→ 39	RCL 5		
→ 15	RCL 6		40	2		
16	STO 2		41	+		
17	PAUSE		42	$\sqrt{x}$		
18	RCL 1		43	STO 6		
19	−		44	GTO 06		
20	RCL 1		45			
21	RCL 0		46			
22	−		47			
23	STO 3		48			
24	−		49			

Program locations 39 through 49 are reserved for the program computing $f(x)$. This program should assume x in R_5 and should put $f(x)$ into R_6. Only the stack may be used for temporary storage. The last instruction should be GTO 06. The program for computing the f of example $\boxed{1}$ is shown above.

5. Operating Instructions

Load the program (including the program to compute f); switch to RUN. Select the mode of displaying numbers, for instance

FIX 8

If f contains trigonometric functions with argument in radians, press

RAD

Load the starting value x_0 into R_0. When

PRGM

R/S

is pressed, the computation starts. The calculator will briefly display each x_{n+2} and stop at display of x_n'. Press

R/S

to start a new cycle. No convergence test is provided nor are iterations counted.

6. Examples and Timing

[1] $f(x) := \sqrt{2 + x}$. Sequence $\{x_n'\}$, if $x_0 = 0$:

$$x_0' = 2.03942606$$
$$x_1' = 2.00208254$$
$$x_2' = 2.00012537$$
$$x_3' = 2.00000776$$
$$x_4' = 2.00000048$$
$$x_5' = 2.00000003$$
$$x_6' = 2.00000000$$

The fixed point $s = 2$ has now been reached in six iterations (which requires the computation of 8 iterates x_n). Computing time is about the same as before, because of the additional work required to compute x_n'.

[2] $f(x) := e^{-x}$. With $x_0 = 0$ we get

$x_2 = 0.36787944$	$x_0' = 0.61269984$
0.69220063	0.58222610
0.50047350	0.57170577

$$x_5 = 0.60624354 \qquad x_3' = 0.56863881$$

$$\cdot \quad \cdot \quad \cdot \quad \cdot \quad \cdot$$

$$x_{16} = 0.56706790 \qquad x_{14}' = 0.56714330$$

$$x_{17} = 0.56718605 \qquad x_{15}' = 0.56714329$$

At this point the sequence $\{x_n'\}$ has converged, whereas the sequence $\{x_n\}$ still has only four correct digits. Computing time about 80 sec.

AITKEN-STEFFENSEN ITERATION

1. Purpose

To compute the fixed points of smooth functions f
without restrictions on the slope of f.

2. Method

Aitken-Steffensen iteration. This is a variant of the
Aitken acceleration considered in the preceding program. Each x_0' is used as a starting point for a new
iteration. Equivalently, we generate a sequence
$\{x^{(n)}\}$ by choosing $x^{(0)}$ arbitrarily and forming

$$x^{(n+1)} := x^{(n)} - \frac{\left[f(x^{(n)}) - x^{(n)}\right]^2}{f(f(x^{(n)})) - 2f(x^{(n)}) + x^{(n)}} \, .$$

If f has a fixed point s, if $f'(s) \neq 1$, and if $x^{(0)}$
is chosen sufficiently close to s, then $x^{(n)} \to s$ with
quadratic convergence (see ENA, § 4.11). If $f'(s) = 1$,
convergence is dubious; in particular, some $x^{(n)}$ may
fail to be defined because of vanishing denominator.

3. Flow Diagram

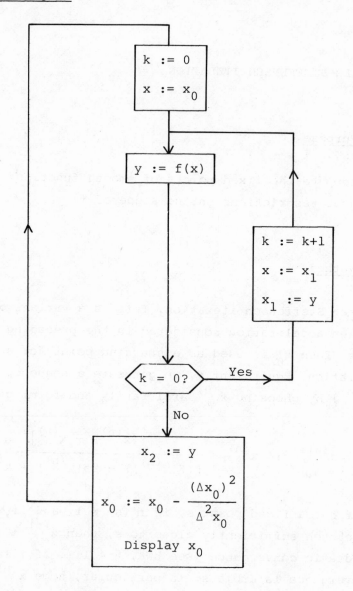

4. <u>Storage and Program</u>

R_0	R_1	R_2	R_3	R_4	R_5	R_6	R_7
$\boxed{x_0}$	x_1	x_2	Δx_0	$\Delta^2 x_0$	x	y	k

00			25	RCL 3	
→ 01	CLX		26	x^2	
02	STO 7		27	RCL 4	
03	RCL 0		28	÷	
04	STO 5		29	CHS	
05	GTO 35		30	RCL 0	
→ 06	RCL 7		31	+	
07	x ≠ 0		32	STO 0	
08	GTO 15		33	R/S	
09	RCL 6		34	GTO 01	
10	STO 1		→ 35	RCL 5	
11	STO 5		36	1	
12	1		37	0	
13	STO+7		38	*	
14	GTO 35		39	CHS	
→ 15	RCL 6		40	e^x	
16	STO 2		41	STO 6	
17	RCL 1		42	GTO 06	
18	-		43		
19	RCL 1		44		
20	RCL 0		45		
21	-		46		
22	STO 3		47		
23	-		48		
24	STO 4		49		

The program to compute f should be in locations 35 through 49. The value of x is in R_5; $y = f(x)$ should be put in R_6. Use the stack only for temporary storage. The last instruction is GTO 06. The program for example 4 is shown above.

5. Operating Instructions

Similar to preceding program. After

PRGM

R/S

is pressed, the computation starts. The computer displays each $x^{(n)}$ and stops. Press

R/S

to start new cycle. No convergence test is provided.

6. Examples and Timing

1    $f(x) := \sqrt{2 + x}$. $x^{(0)} = 0$ yields sequence $\{x^{(n)}\}$ as follows:

$$x^{(1)} = 2.03942606$$
$$x^{(2)} = 2.00000802$$
$$x^{(3)} = 2.00000000$$

The limit has been reached (to within machine accuracy) in three iterations. Computing time 10 sec.

$\boxed{2}$ $f(x) := \dfrac{1}{1 + x}$. $x^{(0)} = 0$ yields

$$x^{(1)} = 0.66666667$$
$$x^{(2)} = 0.61818182$$
$$x^{(3)} = 0.61803399$$

Computing time 15 sec.

$\boxed{3}$ $f(x) := (x - 1)^2$. $x^{(0)} = 3$ yields

$$x^{(1)} = 2.75000000$$
$$2.63888889$$
$$2.61864329$$
$$2.61803453$$
$$2.61803399$$

The last value is $\frac{1}{2}(3 + \sqrt{5})$, the larger fixed point of f. Starting with $x^{(0)} = 0$ we get

$$0.50000000$$
$$0.38888889$$
$$0.38199234$$
$$0.38196601$$
$$0.38196601$$

The last value is $\frac{1}{2}(3 - \sqrt{5})$, the smaller fixed point. Computing time in both cases about 15 sec.

4     $f(x) := e^{-10x}$. With $x^{(0)} = 0$ we get

$$x^{(1)} = 0.50001135$$
$$0.32882657$$
$$0.23868709$$
$$0.19120788$$
$$0.17597093$$
$$0.17456388$$
$$0.17455280$$
$$0.17455280$$

Iteration converges after eight steps. Computing time about 30 sec.

NEWTON ITERATION FOR COMPLEX ROOTS

1. Purpose

To demonstrate the convergence of Newton's method to compute the square root of a given complex number $c = a + ib \neq 0$. This method consists in forming the sequence $\{z_n\} = \{x_n + iy_n\}$ by choosing z_0 and then computing recursively

$$z_{n+1} = \frac{1}{2}(z_n + \frac{c}{z_n}) \ , \quad n = 0, 1, 2, \ldots \ . \qquad (1)$$

This sequence converges to the value of $\sqrt{c}$ nearest to z_0 if there is precisely one such value. It diverges if z_0 lies on the straight line through O perpendicular to the straightline segment joining the two values of $\sqrt{c}$ (see ACCA I, § 6.12).

2. Method

We evaluate the sequence (1) until $|z_{n+1} - z_n| < \epsilon$, a prescribed tolerance. The number of iteration steps is counted.

3. <u>Flow Diagram</u>

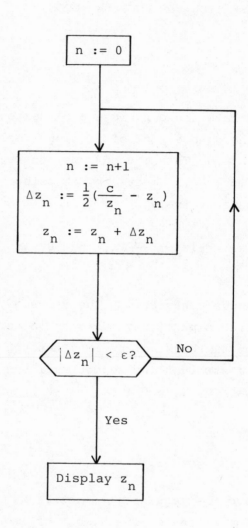

4. Storage and Program

R_0	R_1	R_2	R_3	R_4	R_5	R_6	R_7
a	b	x_0	y_0	ε	$\|z_n\|^2$	Δx_n	n

00			25	RCL 2
01	CLX		26	*
02	STO 7		27	RCL 0
→ 03	1		28	RCL 3
04	STO+7		29	*
05	RCL 3		30	−
06	RCL 2		31	RCL 5
07	→ P		32	÷
08	x^2		33	RCL 3
09	STO 5		34	−
10	RCL 0		35	2
11	RCL 2		36	÷
12	*		37	STO+3
13	RCL 1		38	RCL 6
14	RCL 3		39	STO+2
15	*		40	→ P
16	+		41	STO 6
17	RCL 5		42	RCL 2
18	÷		43	PAUSE
19	RCL 2		44	RCL 4
20	−		45	RCL 6
21	2		46	$x \overset{\geq}{=} y$
22	÷		47	GTO 03
23	STO 6		48	RCL 3
24	RCL 1		49	RCL 2

5. Operating Instructions

Load the program. Turn the operating switch to RUN.
Select the mode of display numbers, for instance

SCI 8

Load the data as follows:

$$
\begin{array}{lll}
a & \text{into} & R_0 \\
b & \text{into} & R_1 \\
x_0 & \text{into} & R_2 \\
y_0 & \text{into} & R_3
\end{array}
$$

Load ε into R_4. For instance, for $\varepsilon = 10^{-9}$ press

EEX

CHS

9

STO 4

When

PRGM

R/S

is pressed, the calculator starts iterating, pausing
briefly after each cycle while displaying x_n. After
convergence has been achieved, the calculator will

stop and display final x_n. To find y_n, press

$$x \overset{>}{<} y$$

To display number of iterations required, press

RCL 7

If all iterates are to be recorded, instruction 43 should be changed to R/S. The calculator will then stop after each cycle, displaying x_n. To display y_n, press

RCL 3

To compute next iterate, press

R/S

6. Examples and Timing

[1] $c = 3 + 4i$, $z_0 = 1$, $\varepsilon = 10^{-9}$. The successive iterates are

$$2.00000000 + 2.00000000 \ i$$
$$1.87500000 + 1.12500000 \ i$$
$$1.99632353 + 0.99387255 \ i$$
$$1.99999968 + 1.00001144 \ i$$
$$2.00000000 + 1.00000000 \ i$$

Convergence has been achieved. Computing time 35 sec.

[2]   c and ε as before, $z_0 = -i$. Algorithm converges to

$$- 2.00000000 - 1.00000000 \, i$$

in $n = 4$ iterations. Computing time 40 sec.

[3] $c = -1 + 0.00001 \, i$, $z_0 = 1$, $\varepsilon = 10^{-9}$. Algorithm converges to

$$0.00000500 + 1.00000000 \, i$$

in $n = 24$ iterations. Computing time about 2 min.

[4] $c = -1$, $z_0 = 1$. No convergence.

Part 3
POLYNOMIALS

HORNER ALGORITHM

1. Purpose

Given an arbitrary polynomial of degree $\leq$ 4 with real
coefficients

$$p(x) = a_0 x^4 + a_1 x^3 + a_2 x^2 + a_3 x + a_4 \ ,$$

and given an arbitrary real number x_0, to determine
the coefficients b_m in the representation of p in
powers of $h := x - x_0$,

$$p(x_0 + h) = b_0 h^4 + b_1 h^3 + b_2 h^2 + b_3 h + b_4 \ ,$$

that is, the Taylor coefficients of p at x_0.

2. Method

The Horner algorithm (see ENA § 3.4; ACCA I § 6.1).
We generate coefficients $b_n^{(m)}$ as follows. Let

$$b_n^{(-1)} := a_n \ , \ n = 0, 1, \ldots , 4 \ ,$$

and for m = 0, 1, ... , 4:

$$b_0^{(m)} := b_0^{(m-1)} \; ; \; b_n^{(m)} := x_0 b_{n-1}^{(m)} + b_n^{(m-1)} \; , \; n = 1, \ldots, 4-m.$$

Then

$$b_m = b_m^{(4-m)} \; , \; m = 0, 1, \ldots , 4.$$

3. Flow Diagram

Because an address modification is not available, the programming here is different from what it would be on a larger computer. By a cyclic permutation (here called rotation) of the a_i, the coefficients operated on are always found in the same registers. After the algorithm is executed, the b_i are stored where, previously, the a_i were stored.

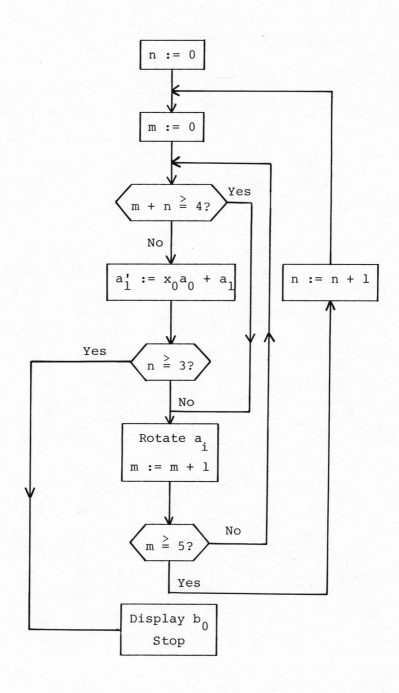

4. Storage and Program

R_0	R_1	R_2	R_3	R_4	R_5	R_6	R_7
a_0	a_1	a_2	a_3	a_4	x_0	temp	n.m

#		#	
00		25	RCL 0
01	CLX	26	STO 6
02	STO 7	27	RCL 1
→ 03	RCL 7	28	STO 0
04	INT	29	RCL 2
05	STO 7	30	STO 1
→ 06	RCL 7	31	RCL 3
07	FRAC	32	STO 2
08	9	33	RCL 4
09	*	34	STO 3
10	RCL 7	35	RCL 6
11	+	36	STO 4
12	3	37	.
13	x < y	38	1
14	GTO 25	39	STO +7
15	RCL 0	40	.
16	RCL 5	41	5
17	*	42	RCL 7
18	RCL 1	43	FRAC
19	+	44	x < y
20	STO 1	45	GTO 06
21	3	46	1
22	RCL 7	47	STO +7
23	$x \overset{>}{=} y$	48	GTO 03
24	GTO 49	→ 49	RCL 0

Note: A fractional index n.m is used to save storage.

5. Operating Instructions

Load the program and switch to RUN. Load the coefficients a_n into R_n ($n = 0, 1, ..., 4$) and x_0 into R_5. When

PRGM

R/S

is pressed, the calculator computes the b_m and stops by displaying b_0. At the same time, the b_m are stored in R_m ($m = 0, 1, ..., 4$), thus enabling the operator to continue the computation immediately with a different x_0. The original a_n are lost.

6. Examples and Timing

$\boxed{1}$ Consider the polynomial

$$p(x) = x^4 - 4x^3 + 3x^2 - 2x + 5 .$$

Carrying through the algorithm with $x_0 = 0.5$ yields the Taylor coefficients at 0.5,

$b_m = 1.0000,   -2.0000,   -1.5000,   -1.5000,   4.3125$

(time required about 29 sec). The last coefficient is $p(0.5)$. Repeating the algorithm with the foregoing data yields the Taylor coefficients at 1,

b_m = 1.0000,   0.0000,   -3.0000,   -4.0000,   3.0000

Repeating three times with x_0 = -0.33333333
yields the coefficient arrays

1.0000,   -1.3333,   -2.3333,   -2.1481,   4.0123
1.0000,   -2.6667,   -0.3333,   -1.1852,   4.5309
1.0000,   -4.0000,   3.0000,   -2.0000,   5.0000

The last array is identical with the initial
array of coefficients, demonstrating the "group
property" of the Horner algorithm.

2   For the polynomial $p(x) = x^4$, $x_0 = 1$ we get

b_m = 1.0000,   4.0000,   6.0000,   4.0000,   1.0000,

the binomial coefficients $\binom{4}{m}$. Computing time
29 sec.

NEWTON'S METHOD FOR POLYNOMIALS

1. Purpose

To determine all real zeros of a real polynomial of degree 4,

$$p(x) = a_0x^4 + a_1x^3 + a_2x^2 + a_3x + a_4 \ . \qquad (1)$$

2. Method

Newton's method, using Horner's scheme to evaluate p and p' and deflating the polynomial after each zero has been found. Newton's method requires forming the sequence $\{x_n\}$ according to

$$x_{n+1} := x_n - \frac{p(x_n)}{p'(x_n)} \ , \qquad (2)$$

with an arbitrarily chosen starting value x_0. The sequence $\{x_n\}$ will converge to a zero, provided a real zero exists and x_0 is chosen close enough to the zero. [Unless $p'(0) = 0$, the choice $x_0 = 0$ will frequently work and will produce convergence to the zero of smallest modulus.] For the purpose of this algorithm, Horner's scheme may be formulated as follows: Compute the array $\{b_i\}$ by

$$b_{-1} := 0, \quad b_k := xb_{k-1} + a_k \ , \quad k = 0, 1, 2, 3, 4;$$

then compute the array $\{c_i\}$ by

$$c_{-1} := 0, \quad c_k := xc_{k-1} + b_k \ , \quad k = 0, 1, 2, 3.$$

Then $p(x) = b_4$ and $p'(x) = c_3$. The iteration (2) is terminated if

$$|b_4| = |p(x_n)| < 10^{-6} \ .$$

(This tolerance may be changed to any one-digit power of 10, but too stiff a tolerance may cause the sequence $\{x_n\}$ to cycle instead of to converge.) If $p(z) = 0$, the b_i are simply the coefficients of the deflated polynomial

$$P_1(x) := \frac{p(x)}{x - z} = b_0x^3 + b_1x^2 + b_2x + b_3 \ .$$

Because the coefficients b_i are available, the deflation thus is achieved by moving the b_i into the locations of the a_i (see operating instructions).

3. <u>Flow Diagram</u>

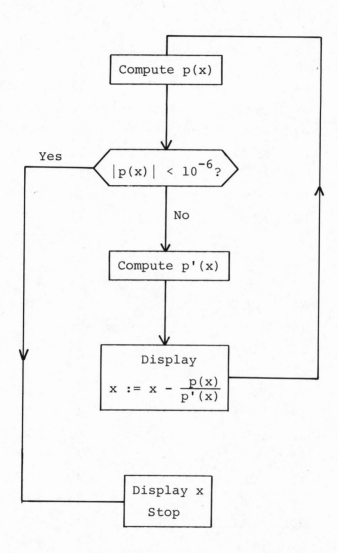

The program presents a particular challenge because
there are ten essential quantities a_0, a_1, a_2, a_3,
a_4, b_1, b_2, b_3, b_4, x that have to be saved.

4. Storage and Program

R_0	R_1	R_2	R_3	R_4	R_5	R_6	R_7
a_0	a_1	a_2	a_3	a_4	b_1	b_2	b_3
					x		

00		25	GTO 48
01	ENTER	26	CLX
02	ENTER	27	LAST X
03	ENTER	28	R ↓
04	RCL 0	29	R ↓
05	*	30	RCL 0
06	RCL 1	31	*
07	+	32	RCL 5
08	STO 5	33	+
09	*	34	$x \lessgtr y$
10	RCL 2	35	STO 5
11	+	36	*
12	STO 6	37	RCL 6
13	*	38	+
14	RCL 3	39	RCL 5
15	+	40	*
16	STO 7	41	RCL 7
17	*	42	+
18	RCL 4	43	÷
19	+	44	STO-5
20	ABS	45	RCL 5
21	EEX	46	PAUSE
22	CHS	47	GTO 01
23	6	48	R ↓
24	$x \geqq y$	49	R ↓

5. Operating Instructions

Load the program. Turn the operating switch to RUN.
Select the mode of displaying numbers, for instance,
by pressing

FIX 8

Load the coefficients a_i of given polynomial into
R_i, i = 0, ... , 4. [Be sure to use indexing of coefficients as in (1).] Load the starting value x_0
into the X register. Press

PRGM

R/S

to start computation. The iterates x_n will be computed; each x_n will be displayed briefly. The calcula-
tor will stop if x_n meets convergence test $|p(x_n)| <$
10^{-6}, displaying final value of x_n. Exponent -6 in
convergence test may be changed to -m by changing instruction 23 to m, where m is any one-digit integer.

   To deflate polynomial when zero has been found,
press

RCL 0

STO 1

RCL 5

STO 2

RCL 6

STO 3

RCL 7

STO 4

CLX

STO 0

Load the starting value x_0 into the X register (if $x_0 = 0$, it is already there) and press

R/S

to restart computation. The process may be repeated until all real zeros have been found.

6. Examples and Timing

$\boxed{1}$ $p(x) := x^4 - 16x^3 + 72x^2 - 96x + 24.$
Starting with $x_0 = 0$ we get the smallest zero

$$z_1 = 0.32254769$$

after four iterations. Deflating and again starting with $x_0 = 0$ yields

$$z_2 = 1.74576110$$

after five iterations. Deflating and starting

with $x_0 = 0$ after five iterations yields

$$z_3 = 4.53662030$$

Deflating once more yields

$$z_4 = 9.39507092$$

in one iteration. Computing time about 4 sec per iteration.

2 $p(x) := x^4 - 10x^3 + 35x^2 - 50x + 24$
$\qquad = (x - 4)(x - 3)(x - 2)(x - 1).$
The program correctly finds the zeros

$$z_1 = 1.00000000$$
$$z_2 = 2.00000000$$
$$z_3 = 3.00000000$$
$$z_4 = 4.00000000$$

Total elapsed time (including deflations) 2.75 min.

BERNOULLI'S METHOD FOR SINGLE DOMINANT ZERO

1. Purpose

To determine the zero of greatest absolute value of a real polynomial of degree 4,

$$p(x) = x^4 + a_1 x^3 + a_2 x^2 + a_3 x + a_4 , \qquad (1)$$

if there is a single such zero.

2. Method

Bernoulli's method (see ENA, § 7.1; ACCA I, § 7.4). A sequence $\{x_n\}$ is generated by choosing starting values x_0, x_1, x_2, x_3 "arbitrarily" and letting

$$x_n := - (a_1 x_{n-1} + a_2 x_{n-2} + a_3 x_{n-3} + a_4 x_{n-4}) ,$$

$n = 4, 5, \ldots$. One forms the quotients

$$q_n := \frac{x_n}{x_{n-1}} ; \qquad (2)$$

if p has a single dominant zero w and if the starting values are not in a certain set of measure zero, then the q_n are defined, at least for sufficiently large n, and

$$w = \lim_{n \to \infty} q_n .$$

In this program the calculation is stopped if

$$\left| \frac{q_n}{q_{n-1}} - 1 \right| < 10^{-8} . \qquad (3)$$

This may cause accidental termination if two consecutive q's happen to be identical. The program will also halt if an x_n becomes accidentally zero, or if there is overflow or underflow because x_n exceeds the range of the computer.

Experience has shown that for polynomials with small integral coefficients these accidents most often are caused by a too special choice of the starting values, such as $x_0 = x_1 = x_2 = x_3 = 1$. The program, therefore, generates its own highly irrational starting values.

3. Flow Diagram

A difficulty is generated by the fact that all available storage locations are taken up by the four co-efficients of the polynomial and by the four currently needed x_n. Therefore, there is no orderly transition from n to n + 1; rather, each x_n is overwritten with x_{n+1} as soon as it is no longer needed. Furthermore, the quotients q_n and q_{n-1} are recalculated each time they are needed.

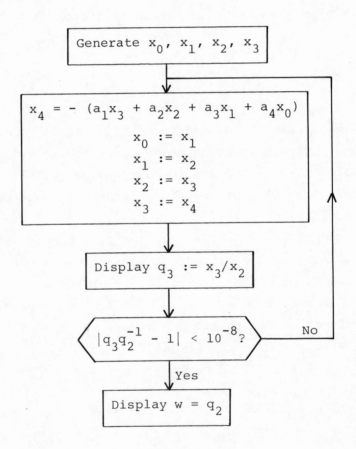

4. Storage and Program

R_0	R_1	R_2	R_3	R_4	R_5	R_6	R_7
a_1	a_2	a_3	a_4	x_3	x_2	x_1	x_0

	00		25	*
	01	π	26	+
	02	STO 4	27	CHS
	03	$\sqrt{x}$	28	STO 4
	04	STO 5	29	RCL 5
	05	e^x	30	$\div$
	06	STO 6	31	PAUSE (or R/S)
	07	log	32	ENTER
	08	STO 7	33	ENTER
$\rightarrow$	09	RCL 3	34	RCL 5
	10	RCL 7	35	RCL 6
	11	*	36	$\div$
	12	RCL 2	37	$\div$
	13	RCL 6	38	1
	14	STO 7	39	–
	15	*	40	ABS
	16	+	41	1
	17	RCL 1	42	EEX
	18	RCL 5	43	CHS
	19	STO 6	44	8
	20	*	45	x < y
	21	+	46	GTO 09
	22	RCL 0	47	R ↓
	23	RCL 4	48	R ↓
	24	STO 5	49	GTO 00

5. Operating Instructions

Load the program, switch to RUN, and choose mode of displaying numbers desired, for instance, by pressing

$$SCI \quad 8$$

to get floating eight-digit display. Another possibility is initially to display few digits only, which makes it easier to check convergence "by eye." Be sure to define coefficients of polynomial as in (1). Load the coefficient a_k into R_{k-1} (k = 1, 2, 3, 4). Press

$$PRGM$$
$$R/S$$

to start computation. The calculator will display briefly each q_n. (If an indeterminate stop is desired, instruction 31 should be changed to R/S. In this case, R/S must be pressed after each display to continue computation.) The calculation will terminate if condition (3) is satisfied; the last q_n will be displayed.

If starting values other than those provided by the program are desired, these should be loaded as follows:

$$
\begin{array}{lll}
x_0 & \text{into} & R_7 \\
x_1 & \text{into} & R_6 \\
x_2 & \text{into} & R_5 \\
x_3 & \text{into} & R_4 \ .
\end{array}
$$

In this case, the calculation is started by pressing

$$GTO\ 09$$
$$R/S$$

6. Examples and Timing

$\boxed{1}$ $p(x) := x^4 - 10x^3 + 35x^2 - 50x + 24$
$$= (x - 4)(x - 3)(x - 2)(x - 1).$$
The calculator displays

$$w = 4.0000000$$

after 60 iterations. Computing time approximately 4 min.

$\boxed{2}$ $p(x) = x^4 + 3x^3 - 7x^2 - 15x + 18$
$$= (x + 3)^2(x - 1)(x - 2).$$
Convergence is very slow because the dominant zero has multiplicity > 1. Faster convergence is achieved (see ENA, § 7.4) by choosing the starting values

$$x_0 = -a_1 = -3$$
$$x_1 = -(2a_2 + a_1x_0) = 23$$
$$x_2 = -(3a_3 + a_2x_0 + a_1x_1) = -45$$
$$x_3 = -(4a_4 + a_3x_0 + a_2x_1 + a_1x_2) = 179\ .$$

This yields

$$w = -2.9999999$$

after 45 iterations. Computing time approximately
3 min.

$\boxed{3}$ $p(x) := x^4 - x^3 - 9x^2 - 2x - 3$ yields

$$w = 3.6652758 \ .$$

Computing time 3.4 min.

BERNOULLI'S METHOD FOR COMPLEX
CONJUGATE DOMINANT ZEROS

1. Purpose

To determine the quadratic factor formed by the two
zeros of largest modulus of a real polynomial of degree 4,

$$p(x) = x^4 + a_1 x^3 + a_2 x^2 + a_3 x + a_4 \ , \qquad (1)$$

provided that all remaining zeros have smaller modulus.

2. Method

Bernoulli's method (see ENA, § 7.5), which in this
situation is identical with a pristine version of the
quotient-difference algorithm (see ACCA I, § 7.6). We
generate the sequence $\{x_n\}$ as solution of the difference equation

$$x_n := - (a_1 x_{n-1} + a_2 x_{n-2} + a_3 x_{n-3} + a_4 x_{n-4}) \ ,$$

$n = 4$, using arbitrarily chosen initial values x_0, x_1, x_2, x_3. With the x_n we form the quotients

$$q_n := \frac{x_n}{x_{n-1}} ,$$

as in the preceding algorithm, and with the quotients the differences

$$e_n := q_n - q_{n-1}$$

and the modified quotients

$$q'_{n-1} := \frac{e_n}{e_{n-1}} q_{n-1} .$$

Finally, we form the quantities

$$r_n := q_{n-1} + q'_{n-1} , \quad s_n := q_{n-2} q'_{n-1} .$$

If p has two dominant zeros w_1 and w_2, and if all other zeros have smaller modulus, then the limits

$$r := \lim_{n \to \infty} r_n , \quad s := \lim_{n \to \infty} s_n$$

exist and satisfy

$$w_1 + w_2 = r \, , \quad w_1 w_2 = s \, ;$$

in other words, w_1 and w_2 may be determined as solutions of the quadratic equation

$$x^2 - rx + s = 0 \, . \tag{2}$$

Our program generates the values r_n and s_n, but it does not check for convergence because of lack of programming space. For the same reason, the starting values x_0, x_1, x_2, x_3 must be supplied by the user.

3. Flow Diagram

Again, there is the difficulty that all available storage is taken up by the coefficients a_k and by the currently needed x_n. Auxiliary quantities, if they cannot be stored in the stack, are recalculated each time they are needed. The x_n are overwritten with x_{n+1} as soon as they are no longer required.

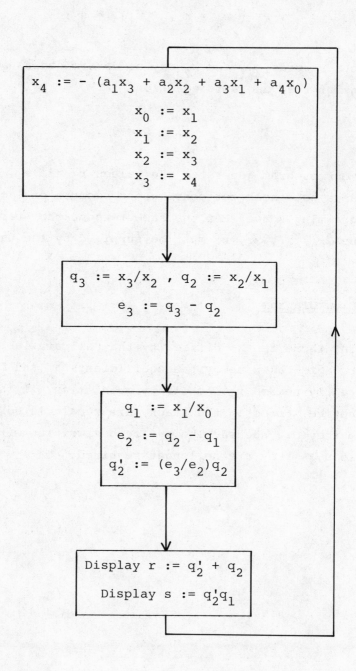

4. Storage and Program

R_0	R_1	R_2	R_3	R_4	R_5	R_6	R_7
a_1	a_2	a_3	a_4	x_3	x_2	x_1	x_0

	00		25	$\div$
$\rightarrow$	01	RCL 3	26	$-$
	02	RCL 7	27	RCL 5
	03	$*$	28	RCL 6
	04	RCL 2	29	$\div$
	05	RCL 6	30	RCL 6
	06	STO 7	31	RCL 7
	07	$*$	32	$\div$
	08	$+$	33	$-$
	09	RCL 1	34	$\div$
	10	RCL 5	35	RCL 5
	11	STO 6	36	RCL 6
	12	$*$	37	$\div$
	13	$+$	38	$*$
	14	RCL 0	39	LAST X
	15	RCL 4	40	$x \lessgtr y$
	16	STO 5	41	$+$
	17	$*$	42	PAUSE
	18	$+$	43	LAST X
	19	CHS	44	RCL 6
	20	STO 4	45	RCL 7
	21	RCL 5	46	$\div$
	22	$\div$	47	$*$
	23	RCL 5	48	R/S
	24	RCL 6	49	GTO 01

5. Operating Instructions

Load the program; switch to RUN. Select the mode of displaying numbers (e.g., SCI 8). Load coefficients a_k of given polynomial into R_{k-1} ($k = 1, \ldots, 4$), being sure to use notation (1). Load irrational starting values (such as $\sqrt{\pi}$, e^{π}, $\sqrt{2}$, etc.) into $R_4, \ldots, R_7$. Press

PRGM

R/S

to start the computation. The calculator will alternate between a brief display of r_n and an indeterminate display of s_n. After each display of s_n,

R/S

must be pressed to start new round of computation. Convergence of the sequences $\{r_n\}$ and $\{s_n\}$ must be tested by eye.

If indeterminate displays of both r_n and s_n are desired, instruction 42 must be changed to R/S. If brief displays of both r_n and s_n are desired, instruction 48 must be changed to PAUSE.

An error halt may occur if an x_n or an e_n becomes accidentally zero, or if there is overflow in x_n. In the first case, the calculation should be repeated, using different starting values. In case of an overflow, the exponents in all x_i should be scaled down by the same number.

6. Examples and Timing

[1] Let $p(x) = x^4 + 3x^3 + 10x^2 - 11x + 7$. We first
use the display mode FIX 2 (for easier reading)
and set both instructions 42 and 48 to R/S. Using
the starting values

$$x_3 := \pi \qquad \to R_4$$
$$x_2 := \sqrt{\pi} \qquad \to R_5$$
$$x_1 := e^{\sqrt{\pi}} \qquad \to R_6$$
$$x_0 := \tan e^{\sqrt{\pi}} \qquad \to R_7$$

we get

r_n	s_n
15.18	4.04
- 7.05	- 35.37
- 3.43	11.32
- 3.85	13.11
- 4.00	13.23

Because convergence appears to be slow, we change
both instructions 42 and 48 to PAUSE and let the
computation go on. After a while, the sequences
$\{r_n\}$ and $\{s_n\}$ will have converged to within 10^{-2}.
We change display to FIX 5 and continue. After
convergence to within 10^{-5} has been achieved, we

change display to FIX 8. After

$$r_n = - 3.97856125 , \quad s_n = 13.36969365$$

has been reached, values do not change anymore. Dominant zeros of polynomial thus are solutions of

$$x^2 + 3.97856125 \, x + 13.36969365 = 0 ,$$

that is,

$$w_1 = - 1.98928063 + 3.06797266 \, i ,$$
$$w_2 = - 1.98928063 - 3.06797266 \, i .$$

2 $p(x) := x^4 + \frac{4}{5}x^3 + \frac{3}{5}x^2 + \frac{2}{5}x + \frac{1}{5} .$

After many iterations and about 18 min computing time, we get

$$r_n = 0.2756646 , \quad s_n = 0.4788911 .$$

Thus p has the approximate quadratic factor

$$x^2 - 0.2756646 \, x + 0.4788911$$

with the zeros

$$w_{1,2} = 0.1378323 \pm 0.6781544 \, i .$$

QUOTIENT-DIFFERENCE ALGORITHM

1. Purpose

To compute the zeros of a real polynomial of degree 4,

$$p(x) = a_0 x^4 + a_1 x^3 + a_2 x^2 + a_3 x + a_4 \ . \qquad (1)$$

2. Method

The progressive version of the qd algorithm (see ENA, § 8.5; ACCA I, § 7.6). For the present purpose this may be described as follows: One computes a sequence of arrays of numbers

$$\underline{a}_n = q_1^{(n)}, \ e_1^{(n)}, \ q_2^{(n)}, \ e_2^{(n)}, \ q_3^{(n)}, \ e_3^{(n)}, \ q_4^{(n)}$$

by the initial conditions

$$q_1^{(0)} := - \frac{a_1}{a_0} \ , \qquad q_2^{(0)} = q_3^{(0)} = q_4^{(0)} = 0 \ ; \qquad (2)$$

$$e_k^{(0)} := \frac{a_{k+1}}{a_k} \ , \ k = 1, \ 2, \ 3, \qquad (3)$$

and by the continuation rules

$$q_k^{(n+1)} := q_k^{(n)} + e_k^{(n)} - e_{k-1}^{(n)} \quad , \quad k = 1,\ 2,\ 3,\ 4, \quad (4)$$

where always $e_0^{(n)} = e_4^{(n)} := 0$, and

$$e_k^{(n+1)} := e_k^{(n)} \cdot \frac{q_{k+1}^{(n+1)}}{q_k^{(n+1)}} \quad , \quad k = 1,\ 2,\ 3. \quad (5)$$

(Note: These formulas are most easily remembered as "rhombus rules":

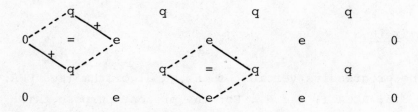

The system of indexing used above is not identical with the indexing used in the sources cited.)

Let the zeros of p be denoted by w_i und numbered such that

$$|w_1| \overset{\geq}{=} |w_2| \overset{\geq}{=} |w_3| \overset{\geq}{=} |w_4| \ .$$

Then under certain conditions - for instance if the w_i are all positive and distinct - all arrays $\underline{a}_n$ exist, and

$$\lim_{n\to\infty} q_k^{(n)} = w_k \; ,$$

$k = 1, 2, 3, 4$. Convergence of $q_k^{(n)}$ to w_k takes place with an error $O(\varepsilon^n)$, where $\varepsilon := \max(|w_{k+1}/w_k|,$ $|w_k/w_{k-1}|)$. By means of auxiliary computations it is also possible to use the arrays $\underline{a}_n$ to compute pairs of complex conjugate zeros (see ACCA I, § 7.9).

3. Flow Diagram

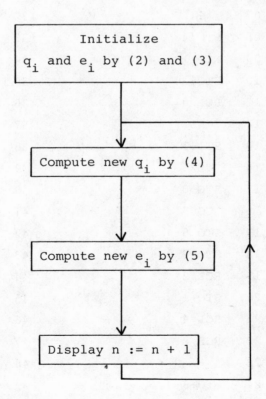

4. Storage and Program

R_0	R_1	R_2	R_3	R_4	R_5	R_6	R_7
a_0	a_1	a_2	a_3	a_4			
n	q_1	e_1	q_2	e_2	q_3	e_3	q_4

00		25	RCL 6
01	RCL 4	26	RCL 4
02	RCL 3	27	−
03	÷	28	STO+5
04	STO 6	29	RCL 6
05	RCL 3	30	STO−7
06	RCL 2	31	RCL 3
07	÷	32	RCL 1
08	STO 4	33	÷
09	RCL 1	34	STO*2
10	STO÷2	35	RCL 5
11	RCL 0	36	RCL 3
12	CHS	37	÷
13	STO÷1	38	STO*4
14	CLX	39	RCL 7
15	STO 0	40	RCL 5
16	STO 3	41	÷
17	STO 5	42	STO*6
18	STO 7	43	1
→ 19	RCL 2	44	RCL 0
20	STO+1	45	+
21	RCL 4	46	STO 0
22	RCL 2	47	PAUSE
23	−	48	GTO 19
24	STO+3	49	

5. Operating Instructions

Load the program, and move the operating switch to RUN. Select the mode of displaying numbers, for instance by pressing

FIX 8

(adequate here because of the convergence of the q's). Load coefficient a_k into R_k (k = 0, 1, ... , 4). Be sure to define coefficients as in (1). Pressing

R/S

will start computation. The calculator will pause briefly after each array is computed, displaying its index n. To inspect array, press

R/S

during pause. The elements of the array computed last are then shown by pressing RCL k (k = 1, ... , 7). Pressing

R/S

will continue the computation.

6. Examples and Timing

$\boxed{1}$ $p(x) = x^4 - 16x^3 + 72x^2 - 96x + 24$, the Laguerre
polynomial of order 4. The initial array (set up
by the program) is

$$16, \ -\frac{72}{16}, \ 0, \ -\frac{96}{72}, \ 0, \ -\frac{24}{96}, \ 0.$$

The program produces the following values $q_k^{(n)}$:

n	$q_1^{(n)}$	$q_2^{(n)}$	$q_3^{(n)}$	$q_4^{(n)}$
10	9.39729639	4.53453197	1.74562397	0.32254767
20	9.39507245	4.53661877	1.74576109	0.32254769
60	9.39507092	4.53662030	1.74576110	0.32254769

Time per array (including pause) approx. 2.75 sec.

$\boxed{2}$ $p(x) := x^4 - 10x^3 + 35x^2 - 50x + 24$
          $= (x - 4)(x - 3)(x - 2)(x - 1)$
Values produced:

n	$q_1^{(n)}$	$q_2^{(n)}$	$q_3^{(n)}$	$q_4^{(n)}$
10	4.0759228	2.9543419	1.9711691	0.9985662
20	4.0040290	2.9965044	1.9994680	0.9999986
30	4.0002261	2.9997832	1.9999908	1.0000000
40	4.0000127	2.9999874	1.9999999	1.0000000
50	4.0000007	2.9999993	2.0000000	1.0000000

ROUTH ALGORITHM

1. Purpose

To determine whether a given real polynomial

$$p(x) := a_0 + a_1 x + a_2 x^2 + \ldots + a_n x^n$$

of degree $n \leq 7$ is <u>stable,</u> that is, whether all its zeros have negative real parts.

2. Method

The Routh algorithm (ACCA I, § 6.7; ACCA II, § 12.7). For the purpose of this program, this algorithm may be described as follows. Let

$$b_k^{(0)} := a_{2k} ,$$

$$b_k^{(1)} := a_{2k+1} ,$$

$k = 0, 1, 2, 3$. (If $n < 7$, $a_{n+1} = \ldots = a_7 := 0$.) For $i = 1, 2, \ldots , n$ do

$$q_i := \frac{b_0^{(i)}}{b_0^{(i-1)}}, \ b_k^{(i+1)} := q_i \ b_{k+1}^{(i-1)} - b_{k+1}^{(i)}, \ k = 0, 1, 2, 3.$$

The polynomial p is stable if and only if the numbers $q_1, q_2, \cdots, q_n$ all exist and are positive.

3. <u>Flow Diagram</u>

We set $q := q_i$, $b_k := b_k^{(i-1)}$, $b_k' := b_k^{(i)}$. To save storage space, quantities b_k are overwritten with b_k' as soon as they are no longer needed.

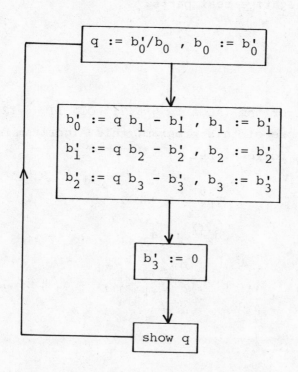

4. Storage and Program

R_0	R_1	R_2	R_3	R_4	R_5	R_6	R_7
a_0	a_1	a_2	a_3	a_4	a_5	a_6	a_7

00		25	RCL 7
→ 01	RCL 0	26	STO 6
02	RCL 1	27	-
03	STO 0	28	STO 5
04	x $\lessgtr$ y	29	CLX
05	÷	30	STO 7
06	ENTER	31	R ↓
07	ENTER	32	R/S
08	ENTER	33	GTO 01
09	RCL 2	34	
10	*	35	
11	RCL 3	36	
12	STO 2	37	
13	-	38	
14	STO 1	39	
15	R ↓	40	
16	RCL 4	41	
17	*	42	
18	RCL 5	43	
19	STO 4	44	
20	-	45	
21	STO 3	46	
22	R ↓	47	
23	RCL 6	48	
24	*	49	

5. Operating Instructions

After loading the program, switch to RUN. Load coefficients a_i of given polynomial into the storage register R_i (i = 0, 1, ... , 7). If degree n is less than
7, the remaining storage registers are to be filled
with 0. Press

PRGM

R/S

The computer will calculate and stop by displaying q_1.
Pressing

R/S

again will result in displaying q_2, and so forth. The
polynomial is stable only if the first n displayed q_i
are positive. The polynomial is unstable if some q_i
< 0 for i $\leq$ n, or if display "Error" indicates divi-
sion by zero.

6. Examples and Timing

$\boxed{1}$ $p(x) = 15 + 22x + 16x^2 + 6x^3 + x^4$. The first
     four q_i are as follows:

$$1.47$$
$$0.79$$
$$0.19$$
$$0.08$$

They are positive; the polynomial thus is stable. Indeed, its zeros are $z = -1 \pm i\sqrt{2}$ and $z = -2 \pm i$.

2 $p(x) = 1 + x + x^2 + x^3 + x^4 + x^5 + x^6 + x^7$. The following output is generated:

$$1.00$$
$$0.00$$
Error

Already $q_2 = 0$, indicating that p is unstable.

3 $p(x) = 6 + x + 4x^2 + x^3 + 3x^4 + 9x^5 + 6x^6 + x^7$. The following q_i's are obtained:

$$0.17$$
$$-\ 0.33$$
$$-\ 24.50$$
$$25.87$$
$$-\ 0.37$$
$$-\ 0.11$$
$$0.11$$

Not all q_i are positive, indicating that p is unstable. Because all q_i exist, the fact that four q_i are negative permits the conclusion that p has precisely four zeros with positive real part (ACCA I, § 6.7).

Total execution time for each of these examples is less than 20 sec.

SCHUR-COHN ALGORITHM I

1. Purpose

To determine whether or not all zeros z_1, z_2, z_3, z_4
of a given polynomial of degree 4,

$$p(x) = a_0 x^4 + a_1 x^3 + a_2 x^2 + a_3 x + a_4 \qquad (1)$$

satisfy $|z_i| < 1$.

2. Method

The Schur-Cohn algorithm (see ACCA I, § 6.8), in
slightly modified form for purposes of efficient computation. For a polynomial of degree n we would form
a triangular array of numbers $a_k^{(m)}$ (m = 0, 1, ... , n;
k = 0, 1, ... , n-m) as follows: Let

$$a_k^{(0)} := a_k , \quad k = 0, 1, ... , n.$$

Then for m = 0, 1, ... , n-1 do

$$q_{m+1} := \frac{a_{n-m}^{(m)}}{a_0^{(m)}} ;$$

$$a_k^{(m+1)} := a_k^{(m)} - q_{m+1} a_{n-m-k}^{(m)} , \quad k = 0, 1, \ldots , n-m-1.$$

All zeros of p satisfy $|z_i| < 1$ if and only if the q_m exist and satisfy

$$|q_m| < 1, \quad m = 1, 2, \ldots , n.$$

3. Flow Diagram

The program implements the algorithm by means of a onedimensional array of quantities b_k ($k = 0, 1, \ldots , n$) which in case of a polynomial of degree n would be defined as follows:

$$b_k := a_0^{(m)} , \quad k = 0, 1, \ldots , m;$$

$$b_k := a_{k-m}^{(m)} , \quad k = m+1, \ldots , n.$$

In terms of the b_k the algorithm performs as follows: For $m = 0, 1, \ldots , n-1$ do

$$q := \frac{b_n}{b_0} , \qquad b_0 := b_0 - qb_n ; \qquad (2)$$

$$b_k' := b_k - qb_n \ , \ k = 1, \ 2, \ \ldots \ , \ m; \tag{3a}$$

$$b_k' := b_k - qb_{n-(k-m)} \ , \ k = m+1, \ \ldots \ , \ n-1; \tag{3b}$$

$$b_k := b_{k-1}' \ , \ k = 1, \ 2, \ \ldots \ , \ n. \tag{4}$$

To carry out (3), the last n-1 elements of the sequence b_1, b_2, $\ldots$, b_{n-1}, b_n, $\ldots$, b_n, where b_n occurs m-1 times, are stored in the stack of the calculator. On the HP-25, this is possible for n = 4. The program is redundant inasmuch b_0 is calculated m+1 times in the m-th cycle.

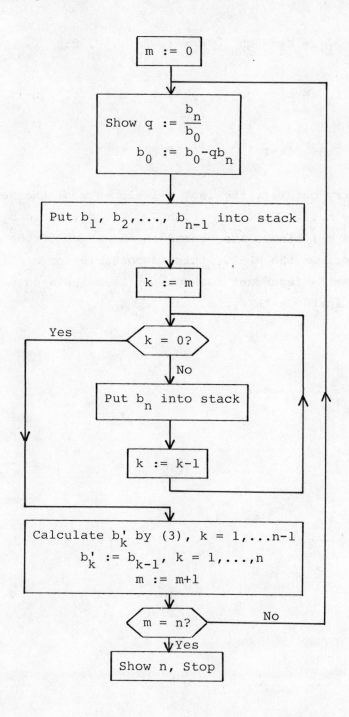

4. Storage and Program

R_0	R_1	R_2	R_3	R_4	R_5	R_6	R_7
a_0	a_1	a_2	a_3	a_4			

	00		25	RCL 5
	01	CLX	26	*
	02	STO 7	27	STO-1
→	03	STO 6	28	R ↓
	04	RCL 4	29	RCL 5
	05	RCL 0	30	*
	06	÷	31	STO-2
	07	STO 5	32	R ↓
	08	PAUSE	33	RCL 5
	09	RCL 4	34	*
	10	*	35	STO-3
	11	STO-0	36	RCL 3
	12	RCL 1	37	STO 4
	13	RCL 2	38	RCL 2
	14	RCL 3	39	STO 3
→	15	RCL 6	40	RCL 1
	16	x = 0	41	STO 2
	17	GTO 24	42	RCL 0
	18	R ↓	43	STO 1
	19	RCL 4	44	1
	20	1	45	STO+7
	21	STO-6	46	4
	22	R ↓	47	RCL 7
	23	GTO 15	48	x < y
	24	R ↓	49	GTO 03

5. Operating Instructions

Load the program. Move the operating switch to RUN. Select the mode of displaying numbers. (Because it only matters whether $|q_i| < 1$ or $|q_i| \geq 1$, FIX 2 is adequate except in doubtful cases.) Load coefficient a_k of given polynomial (1) into R_k (k = 0, 1, ... , 4). When

PRGM

R/S

is pressed, the calculator will carry out the algorithm, briefly displaying each q_k (k = 1, 2, 3, 4), and stop by showing n = 4. The zeros of p all lie inside the unit circle if and only if all q_k exist (no error halt due to division by 0), and if they satisfy $|q_k| < 1$.

If it is desired to determine the number of zeros satisfying $|z_k| > 1$, the quantities q_i should be recorded and used as input for the program "Schur-Cohn Algorithm II". Instruction 08 should be changed to R/S in this case, and R/S should be pressed after display of each q_i.

6. Examples and Timing

[1] $p(x) = 5x^4 + 4x^3 + 3x^2 + 2x + 1$.

The algorithm yields, using FIX 1 display for quick reading,

k	1	2	3	4
q_k	0.2	0.3	0.3	0.5

Computing time 14 sec. All $|q_k| < 1$, thus all zeros of p satisfy $|z_i| < 1$. This could also have been inferred from the Gauss-Lucas theorem (ACCA I, § 6.5), since $p(x) = \dfrac{d}{dx} \dfrac{x^6 - 1}{x - 1}$, or from the Eneström-Kakeya theorem (ACCA I, § 6.4), because the coefficients a_k are positive and from a monotonically decreasing sequence.

2   $p(x) = 4x^4 + 7x^3 + 5x^2 + 3x + 1$.
Using FIX 1 display the program obtains

k	1	2	3	4
q_k	0.3	0.3	0.5	1.0

The value 1.0 is doubtful. We repeat the calculation with FIX 8 display, obtaining

q_k	0.25000000	0.33333333	0.50000000	1.00000000

Not all $|q_k| < 1$, hence not all zeros satisfy $|z_i| < 1$. Indeed, $p(-1) = 0$.

SCHUR-COHN ALGORITHM II

1. Purpose

To determine the number of zeros z_i of a polynomial of degree 4,

$$p(x) = a_0 x^4 + a_1 x^3 + a_2 x^2 + a_3 x + a_4 , \qquad (1)$$

that satisfy $|z_i| > 1$, that is, that lie outside the unit circle.

2. Method

The Schur-Cohn algorithm (see ACCA I, § 6.8). For a polynomial of degree n this runs as follows: If all numbers q_i determined by the Schur-Cohn algorithm I exist and satisfy $|q_i| \neq 1$, the number m of zeros out-side the unit circle equals m_n, where m_n is determined recursively as follows:

$$m_0 := 0 ;$$

$$
m_k := \begin{cases} m_{k-1} & \text{if } |q_{n+1-k}| < 1 \\ \\ k - m_{k-1} & \text{if } |q_{n+1-k}| > 1 \end{cases} \quad , \quad k = 1, 2, \ldots, n. \quad (2)
$$

If some $|q_k| = 1$, the number m cannot be determined by this algorithm. Our program then displays m = 99.

3. Flow Diagram

This is a straightforward realization of the formulas (2). Because the q_k under examination always has to be at the same place, the remaining q_i are shifted up one place at the end of each cycle. The program implements the case n = 4.

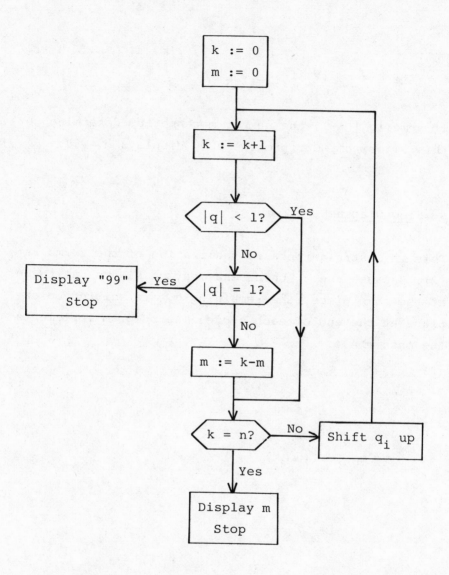

4. Storage and Program

R_0	R_1	R_2	R_3	R_4	R_5	R_6	R_7
	q_1	q_2	q_3	q_4		k	m

00		25	STO 2
01	CLX	26	GTO 04
02	STO 6	→ 27	9
03	STO 7	28	9
→ 04	1	29	GTO 00
05	STO+6	→ 30	RCL 7
06	RCL 4	31	GTO 00
07	ABS	32	
08	x < y	33	
09	GTO 16	34	
10	x = y	35	
11	GTO 27	36	
12	RCL 6	37	
13	RCL 7	38	
14	-	39	
15	STO 7	40	
→ 16	RCL 6	41	
17	4	42	
18	x = y	43	
19	GTO 30	44	
20	RCL 3	45	
21	STO 4	46	
22	RCL 2	47	
23	STO 3	48	
24	RCL 1	49	

5. Operating Instructions

Load the program and move operating switch to RUN.
Press

FIX 0

for integer display of number of zeros. Load q_i into
R_i, i = 1, 2, 3, 4. When

PRGM
R/S

is pressed, the calculator after a short time will
display the number m of zeros outside the unit circle.
If some $|q_i| = 1$ and hence the algorithm fails, the
integer 99 will be displayed.

6. Examples and Timing

1   With the data of example 1 of the preceding
    algorithm we get

m = 0 .

Indeed, we already know that this polynomial has
no zeros outside the unit circle. Computing time
about 4 sec.

2 $p(x) = 27x^4 - 18x^3 + 52x^2 + 3x - 15$.

The program "Schur-Cohn Algorithm I" yields the values

$$q_k = -0.6, \ -0.4, \ 4.7, \ 0.2$$

Using these data, the present algorithm yields $m = 2$, indicating that precisely two zeros of p satisfy $|z_i| > 1$.

Part 4
POWER SERIES

RECIPROCAL POWER SERIES

1. Purpose

Given a formal power series with leading coefficient 1,

$$P = 1 + a_1 x + a_2 x^2 + \ldots \, ,$$

to compute the coefficients b_n of the reciprocal power series

$$P^{-1} = 1 + b_1 x + b_2 x^2 + \ldots$$

satisfying $P^{-1} P = 1$.

2. Method

This is one instance where the very limitations of the pocket calculator make it necessary to invent a new and unorthodox algorithm to solve the problem. Ordinarily one would use the fact that

$$(1 + a_1 x + a_2 x^2 + \ldots)(1 + b_1 x + b_2 x^2 + \ldots) = 1,$$

which implies for n = 1, 2, ...

$$b_n + a_1 b_{n-1} + a_2 b_{n-2} + \ldots + a_n = 0 ,$$

to obtain the recurrence relation $b_0 = 1$,

$$b_n = - a_1 b_{n-1} - a_2 b_{n-2} - \ldots - a_n b_0 ,$$

n = 1, 2, ... However, to compute b_n by this method requires computing and storing all b_k where k < n, which exceeds the storage capacity of the pocket calculator except for trivial values of n. The same is true for "fast" algorithms for computing the reciprocal series using Newton's method and the Fast Fourier Transform (see Aho, Hopcroft, and Ullmann, The design and analysis of computer algorithms, Addison-Wesley, 1974).

Instead, we use the following approach which, however slow the resulting algorithm, works on the HP-25. Writing

$$Q := a_1 x + a_2 x^2 + a_3 x^3 + \ldots ,$$

so that P = 1 + Q, we obviously have

$$P^{-1} = 1 - Q + Q^2 - Q^3 + \ldots .$$

For any $n \geq 1$, the coefficient b_n thus is the sum of all coefficients of the power x^n in all powers Q^k where $1 \leq k \leq n$. In a fixed power Q^k, the coefficient of x^n arises as the sum of all products

$$(-1)^h \, a_{k_1} \, a_{k_2} \, \cdots \, a_{k_h} \, ,$$

where $k_i = 1$, $i = 1, 2, \ldots , h$, and

$$k_1 + k_2 + \ldots + k_h = n \, .$$

Summing over all h, we thus have

$$b_n = \sum (-a_{k_1})(-a_{k_2})\ldots(-a_{k_h}) \, , \qquad (1)$$

where the sum comprises all systems of indices k_1, k_2, $\ldots$, k_h such that $h \geq 1$,

$$k_i \geq 1, \quad i = 1, 2, \ldots , h,$$

$$(2)$$

$$k_1 + k_2 + \ldots + k_h = n.$$

(No upper limit on h needs to be imposed.)

At first sight, it seems difficult to program a complicated formula such as (1) for a pocket calcula-tor, due to the many systems of indices $(k_1, k_2, \ldots, k_h)$ that have to be kept track of. However, the following device, suggested to the author by E. Specker, makes the programming possible. The systems of indices $\underline{k}$ = $(k_1, k_2, \ldots, k_h)$ satisfying (2) are put into a one-to-one correspondence with the integers m satisfying $2^{n-1} \leq m < 2^n$, in the following way: Let

$$m = \varepsilon_0 + 2\varepsilon_1 + 2^2\varepsilon_2 + \ldots + 2^{n-1}\varepsilon_{n-1} \qquad (3)$$

be the binary representation of m (ε_j = 0 or 1, j = 0, 1, ... , n-2; ε_{n-1} = 1). For each j such that ε_j = 1, we put a mark at the point x = j of the real line. In addition, we put a mark at x = 0. These marks divide the interval $[0, n-1]$ into subintervals. The indices k_i of the system $(k_1, k_2, \ldots, k_h)$ corresponding to m are the lengths of these subintervals.

Accordingly, the coefficients b_n may be calculated by the formula

$$b_n = \sum_{m=2^{n-1}}^{2^n-1} \Pi_m \ ,$$

where the products Π_m are defined by means of the binary expansion of m,

$$\Pi_m = (-a_{k_1})(-a_{k_2})\ldots(-a_{k_h}) \ ;$$

here $k_1, k_2, \ldots, k_h$ are the gaps between the non-zero ε_j in (3).

The binary expansion of m is calculated by the following algorithm. Let $p_0 := m$,

$$p_j := INT \frac{p_{j-1}}{2} \, , \, j = 1, \, 2, \, \ldots \, , \, n.$$

Then

$$\varepsilon_j = \begin{cases} 1, \text{ if FRAC } \dfrac{p_j}{2} \neq 0 \, , \\[2em] 0, \text{ otherwise.} \end{cases}$$

Given m, the product $\Pi := \Pi^{(m)}$ thus is calculated re-cursively as follows: Let $\Pi^{(0)} := 1$,

$$\Pi^{(j+1)} := \begin{cases} \Pi^{(j)} & \varepsilon_j = 0 \, , \\[2em] - a_k \, \Pi^{(j)} \, , & \varepsilon_j \neq 0 \, ; \end{cases}$$

here $k := j - j'$ where $j' := \max \{ \, i \mid i < j, \, \varepsilon_i \neq 0 \, \}$. Then $\Pi^{(n)} = \Pi$.

3. Flow Diagram

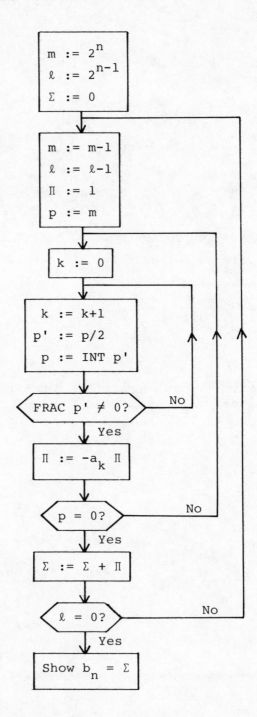

4. Storage and Program

R_0	R_1	R_2	R_3	R_4	R_5	R_6	R_7
	2^n	2^{n-1}	k		p	Π	0
	m	ℓ					Σ

	00		25	
→	01	1	26	
	02	STO 6	27	
	03	STO-1	28	
	04	STO-2	29	
	05	RCL 1	30	
	06	STO 5	31	
→	07	CLX	32	
	08	STO 3	33	
→	09	1	34	
	10	STO+3	35	
	11	RCL 5	36	
	12	2	37	
	13	÷	38	
	14	INT	39	CHS
	15	STO 5	40	STO*6
	16	LAST x	41	RCL 5
	17	FRAC	42	x ≠ 0
	18	x = 0	43	GTO 07
	19	GTO 09	44	RCL 6
	20		45	STO+7
	21		46	RCL 2
	22		47	x ≠ 0
	23		48	GTO 01
	24		49	RCL 7

Locations 20 ÷ 38 are reserved for the program for the computation of a_k. This program should assume k in R_3; at the end of program execution, a_k must be in X register. The registers R_0 and R_4 are available for auxiliary storage.

5. Operating Instructions

Load the program, including the program to compute a_k. Unused locations must be filled up by NOP instructions. Move the operating switch to RUN. Select the mode of displaying numbers. If coefficient b_n is desired, load data as follows:

$$
\begin{array}{rcl}
2^n & \text{into} & R_1 \\
2^{n-1} & \text{into} & R_2 \\
0 & \text{into} & R_7
\end{array}
$$

Here the numbers 2^n and 2^{n-1} must be genuine integers; the calculator will not deal correctly with a 2^n that is computed by the y^x instruction. Press

PRGM

R/S

to start computation. After some time (see examples) the calculator will display b_n. Caution: b_n is calculated as a sum of 2^n terms which may differ in sign and size. Thus already for moderate n, b_n may be affected by rounding error.

6. Examples

[1] To calculate the coefficients b_n in the expansion

$$- \frac{x}{\text{Log}(1 - x)} = 1 + b_1 x + b_2 x^2 + \dots .$$

(The b_n are required for certain methods of numerical integration; see ENA, § 13.4.) The foregoing series is the reciprocal of

$$P := - \frac{\text{Log}(1 - x)}{x} = 1 + \frac{1}{2}x + \frac{1}{3}x^2 + \dots .$$

The coefficients

$$a_k = \frac{1}{k + 1} , \quad k = 1, 2, \dots ,$$

are generated by means of the program

20	RCL 3	24	
21	1	.	
22	+	.	NOP
23	1/x	38	

Results:

n	b_n (calculated)	b_n (exact value)	Computing time
1	-0.500000000	$-\dfrac{1}{2} = -0.500000000$	2 sec
2	-0.083333333	$-\dfrac{1}{12} = -0.833333333$	8 sec
3	-0.041666667	$-\dfrac{1}{24} = -0.041666667$	18 sec
4	-0.026388889	$-\dfrac{19}{720} = -0.026388889$	45 sec
5	-0.018750000	$-\dfrac{3}{160} = -0.018750000$	109 sec
6	-0.014269180	$-\dfrac{863}{60480} = -0.014269180$	252 sec

All calculated values are accurate in all digits given.

2 To compute the Bernoulli numbers B_k defined by

$$\frac{x}{e^x - 1} = \sum_{k=0}^{\infty} \frac{B_k}{k!} x^k$$

(see ACCA I, p. 13). - Setting

$$\frac{e^x - 1}{x} = 1 + a_1 x + a_2 x^2 + \ldots ,$$

the program will calculate the numbers

$$b_n := \frac{B_n}{n!} .$$

The coefficients

$$a_k = \frac{1}{(k+1)!}$$

may be calculated by the following program:

20	1	28	GTO 23
21	STO+3	29	RCL 3
22	RCL 3	30	1/x
23	1	31	
24	−	.	
25	x = 0	.	} NOP
26	GTO 29	.	
27	STO*3	38	

Results:

n	b_n (calculated)	b_n (exact values)	Computing time
1	-0.500000000	$-\frac{1}{2}$ = -0.500000000	3 sec
2	0.083333333	$\frac{1}{12}$ = 0.083333333	8 sec
3	$3 * 10^{-11}$	0	20 sec
4	-0.001388889	$-\frac{1}{720}$ = -0.001388889	54 sec
5	$9 * 10^{-12}$	0	129 sec
6	0.000033069	$\frac{1}{30240}$ = 0.000033069	303 sec

POWER OF POWER SERIES

1. Purpose

Given a formal power series with leading coefficient 1,

$$P = 1 + a_1 x + a_2 x^2 + a_3 x^3 + \dots \ , \qquad (1)$$

and given a real number α, to compute the coefficients $a_n^{(\alpha)}$ of the α-th power of P,

$$P^\alpha = (1 + a_1 x + a_2 x^2 + \dots)^\alpha$$

$$= 1 + a_1^{(\alpha)} x + a_2^{(\alpha)} x^2 + \dots \ .$$

2. Method

By conventional wisdom, the purpose would be served by means of a recurrence relation derived from the fact that the series $R := P^\alpha$ satisfies the formal differential equation

$$P \, R' - \alpha \, P' \, R = 0$$

(J. C. P. Miller formula; see ACCA I, Theorem 1.6c).
This approach is not feasible for pocket calculators
due to storage limitations. Instead, we compute the
series P^α directly from its definition as composition
of the binomial series

$$(1 + x)^\alpha = \sum_{h=0}^{\infty} \frac{(-\alpha)_h}{h!} (-x)^h$$

with the series

$$Q := a_1 x + a_2 x^2 + a_3 x^3 + \ldots .$$

This yields

$$P^\alpha = 1 + \sum_{h=1}^{\infty} \frac{(-\alpha)_h}{h!} (-Q)^h ,$$

hence

$$a_n^{(\alpha)} = \sum \frac{(-\alpha)_h}{h!} (-a_{k_1}) (-a_{k_2}) \ldots (-a_{k_h}) , \qquad (2)$$

where the sum comprises all systems of indices $(k_1, k_2,$
$\ldots, k_h)$ satisfying the conditions (2) of the preceding
algorithm. These systems are coded in terms of the binary representation of the integers m satisfying 2^{n-1}
$\leq m < 2^n$, and we obtain

$$a_n^{(\alpha)} = \sum_{m=2^{n-1}}^{2^n-1} \frac{(-\alpha)_h}{h!} \Pi_m \, , \tag{3}$$

where Π_m has the same meaning as in the preceding algorithm, and where h is the number of the non-zero digits in the binary representation of m.

3. <u>Flow Diagram</u>

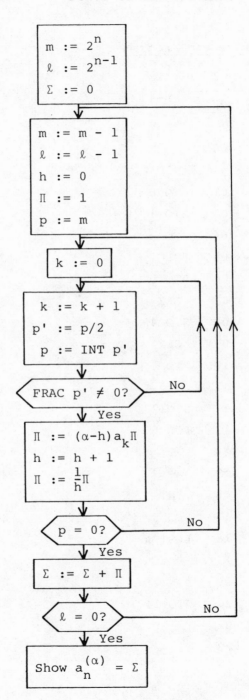

4. Storage and Program

R_0	R_1	R_2	R_3	R_4	R_5	R_6	R_7
α	2^n	2^{n-1}	k	h	p	Π	0
	m	ℓ					Σ

00		25		
→ 01	1	26		
02	STO 6	27		
03	STO−1	28		
04	STO−2	29		
05	RCL 1	30		
06	STO 5	31		
07	CLX	32	STO*6	
08	STO 4	33	RCL 0	
→ 09	CLX	34	RCL 4	
10	STO 3	35	−	
→ 11	1	36	STO*6	
12	STO+3	37	1	
13	RCL 5	38	STO+4	
14	2	39	RCL 4	
15	÷	40	STO÷6	
16	INT	41	RCL 5	
17	STO 5	42	x ≠ 0	
18	LAST x	43	GTO 09	
19	FRAC	44	RCL 6	
20	x = 0	45	STO+7	
21	GTO 11	46	RCL 2	
22		47	x ≠ 0	
23		48	GTO 01	
24		49	RCL 7	

Locations 22 $\div$ 31 are available for the program to compute a_k. This program should assume k in R_3; at the end of program execution, a_k must be in X register. (If it is more convenient to compute $1/a_k$, this may be done; instruction 32 must then be changed to STO$\div$6.) No registers are available for auxiliary storage, but stack may be used freely.

5. Operating Instructions

Load the program, including the program to compute a_k. Unused program locations are to be filled with NOP instructions. Move the operating switch to RUN. Select the mode of displaying numbers. Load exponent

$$\alpha \quad \text{into} \quad R_0$$

and, if coefficient $a_n^{(\alpha)}$ is desired,

$$2^n \quad \text{into} \quad R_1$$
$$2^{n-1} \quad \text{into} \quad R_2$$
$$0 \quad \text{into} \quad R_7$$

(The powers of 2 should be loaded as integers, without using the y^x instruction.) Press

PRGM

R/S

to start computation. The calculator will stop by displaying $a_n^{(\alpha)}$. Caution: Computing time increases exponentially with n. For large n, $a_n^{(\alpha)}$ may be contaminated by rounding error.

6. Examples and Timing

1 Power series reversion. Let y be given by a
power series in x of the form

$$y = x + a_1 x^2 + a_2 x^3 + \ldots := R .$$

Then the expansion of x in terms of y is

$$x = y + c_2 y^2 + c_3 y^3 + \ldots ,$$

where, by the Lagrange-Bürmann Theorem (see ACCA
I, § 1.9)

$$c_n = \frac{1}{n} \text{ res } R^{-n} .$$

Here res R^{-n} denotes the coefficient of x^{-1} in
the expansion of R^{-n}. Because $R = xP$, where P has
the form (1), this coefficient in the foregoing
notation equals $a_{n-1}^{(-n)}$. There follows

$$c_n = \frac{1}{n} a_{n-1}^{(-n)} , \quad n = 2, 3, \ldots .$$

As an example, consider

$$R = e^x - 1 \, ,$$

where

$$a_k = \frac{1}{(k+1)!} \, .$$

Because

$$R^{\boxed{-1}} = \text{Log}(1 + x) = \sum_{n=1}^{\infty} \frac{(-1)^{n-1}}{n} x^n \, ,$$

we have in this case

$$c_n = \frac{(-1)^{n-1}}{n} \, , \qquad a_{n-1}^{(-n)} = (-1)^{n-1} \, .$$

The following program is used to compute the a_k:

22	1	28	GTO 31
23	STO+3	29	STO*3
24	RCL 3	30	GTO 25
25	1	31	RCL 3
26	-	32	STO÷6
27	x = 0		

Being ignorant of the exact answer, the main program then computes the following values:

n	$a_{n-1}^{(-n)}$	Computing time
2	-1.000000000	4 sec
3	1.000000000	10 sec
4	-1.000000000	23 sec
5	0.999999999	58 sec
6	-1.000000000	137 sec
7	0.999999998	5.5 min
8	-1.000000006	12 min
9	1.000000008	
10	-0.999999858	
11	0.999999631	

2 <u>Stirling's formula.</u> The coefficients c_k in the asymptotic expansion of the Γ function,

$$\Gamma(x) \stackrel{\sim}{\sim} \sqrt{2\pi} \ x^{x-1/2} \ e^{-x} \ \{1 + \frac{c_1}{x} + \frac{c_2}{x^2} + \ldots\}, \ x \to \infty,$$

can be defined as

$$c_q := 2^q \ (\tfrac{1}{2})_q \ b_{2q}^{(-q-\tfrac{1}{2})} \ ,$$

where

$$P = \sum_{k=0}^{\infty} a_k x^k = 1 + \frac{2}{3}x + \frac{2}{4}x^2 + \frac{2}{5}x^3 + \ldots$$

(see ACCA II, § 11.6). The program calculates the following values:

q	$b_{2q}^{(-q-\frac{1}{2})}$	c_q	exact value
1	0.083333334	0.083333334	$\frac{1}{12}$ = 0.083333333
2	0.001157408	0.003472223	$\frac{1}{288}$ = 0.003472222
3	-0.000178759	-0.002681385	$-\frac{139}{51840}$ = -0.002681327
4	-0.000002234	-0.000234570	$-\frac{571}{2488320}$ = -0.000229472

EXPONENTIATION OF POWER SERIES

1. Purpose

Given a formal power series with constant coefficient zero,

$$Q = a_1 x + a_2 x^2 + \ldots ,$$

to calculate the coefficients c_n of the series

$$e^Q = 1 + c_1 x + c_2 x^2 + \ldots .$$

2. Method

To use the differential equation satisfied by the series $S := e^Q$ is not feasible due to the lack of storage facilities. However, from the fact that

$$e^Q = 1 + \sum_{h=1}^{\infty} \frac{1}{h!} Q^h$$

we have, using the notation of the two preceding programs,

$$c_n = \frac{1}{h!} \, a_{k_1} \, a_{k_2} \cdots a_{k_h} \, ,$$

where the sum comprises all systems of indices $(k_1, k_2, \ldots, k_h)$ satisfying the conditions (2) of the program "Reciprocal Power Series". As before, these systems are coded in terms of the binary representation of the integers m satisfying $2^{n-1} \leq m < 2^n$, and we thus obtain

$$c_n = \sum_{m=2^{n-1}}^{2^n - 1} \frac{1}{h!} \, \Pi_m \, ,$$

where Π_m has the same meaning as in the two preceding programs.

3. <u>Flow Diagram</u>

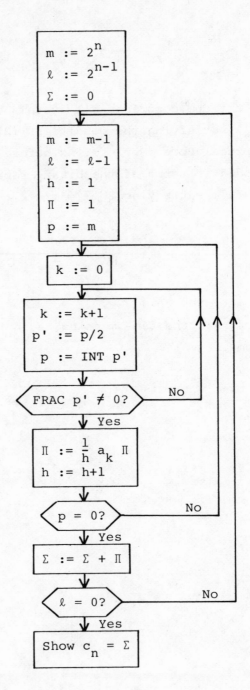

4. Storage and Program

R_0	R_1	R_2	R_3	R_4	R_5	R_6	R_7
	2^n	2^{n-1}	k	h	p	Π	0
	m	ℓ					Σ

00		25	
→ 01	1	26	
02	STO 6	27	
03	STO 4	28	
04	STO-1	29	
05	STO-2	30	
06	RCL 1	31	
07	STO 5	32	
→ 08	CLX	33	
09	STO 3	34	
→ 10	1	35	
11	STO+3	36	STO*6
12	RCL 5	37	RCL 4
13	2	38	STO÷6
14	÷	39	1
15	INT	40	STO+4
16	STO 5	41	RCL 5
17	LAST x	42	x ≠ 0
18	FRAC	43	GTO 08
19	x = 0	44	RCL 6
20	GTO 10	45	STO+7
21		46	RCL 2
22		47	x ≠ 0
23		48	GTO 01
24		49	RCL 7

Locations 21 ÷ 35 are to be used for the program to
compute a_k. This program should assume k in R_3; at the
end of program execution, a_k must be in X register.
(If it is more convenient to compute $1/a_k$, this may be
done; instruction 36 must then be changed to STO÷6.)
Register R_0 is available for auxiliary storage. Unused
locations are to be filled with NOP instructions.

5. Operating Instructions

Load the program, including the program to compute a_k.
Move the operating switch to RUN. Select the mode of
displaying numbers. If coefficient c_n is desired, load

$$2^n \quad \text{into} \quad R_1$$
$$2^{n-1} \quad \text{into} \quad R_2$$
$$0 \quad \text{into} \quad R_7$$

(The powers of 2 should be loaded as integers, without
using the y^x instruction.) Press

PRGM

R/S

to start computation. The calculator will stop by displaying c_n. (Computing time increases exponentially
with n.) Caution: For large n, if some a_k are negative,
cancellation may cause c_n to be contaminated by rounding error.

6. Examples and Timing

1 $Q = \mathrm{Log}(1 - x) = - \sum_{k=1}^{\infty} \frac{1}{k} x^k$.

We should obtain

$$e^Q = 1 - x ,$$

that is, $c_1 = -1$, $c_k = 0$ for $k > 1$. The program to compute a_k is as follows:

$$
\begin{array}{rl}
21 & \text{RCL } 3 \\
22 & \text{CHS} \\
23 & \text{GTO } 36 \\
\left.\begin{array}{r} 24 \\ \vdots \\ 35 \end{array}\right\} & \text{NOP} \\
36 & \text{STO} \div 6
\end{array}
$$

Results:

n	c_n	Computing time
1	-1.000000000	3 sec
2	0.000000000	6 sec
3	0.000000000	15 sec
4	$1 * 10^{-10}$	38 sec
5	0.000000000	90 sec

6	0.000000000	3.0 min
7	$1 * 10^{-10}$	8.1 min

2 The <u>Bell numbers</u> b_n are defined by the expansion

$$e^{e^x - 1} = \sum_{n=0}^{\infty} \frac{b_n}{n!} x^n .$$

For

$$Q = e^x - 1 = \sum_{k=1}^{\infty} \frac{1}{k!} x^k$$

we may calculate a_k by the program

21	RCL 3	28	RCL 3
22	1	29	STO÷6
23	−	30	GTO 37
24	x = 0	31	
25	GTO 28	⋮	NOP
26	STO*3	⋮	
27	GTO 22	36	

Results:

n	c_n	$b_n = n! c_n$	Computing time
1	1.000000000	1	3 sec
2	1.000000000	2	8 sec
3	0.833333333	5	18 sec

4	0.625000000	15	46 sec
5	0.433333333	52	110 sec
6	0.281944445	203	260 sec
7	0.174007996	877	10 min
8	0.102678571	4140	23 min
9	0.058275463	21147	51 min

Part 5

INTEGRATION

NUMERICAL INTEGRATION WITH STEP REFINEMENT

1. Purpose

To evaluate the integral

$$I := \int_0^b f(x)\ dx$$

numerically for an arbitrary function f that can be
evaluated by a program requiring no more than 16 instructions. The general integral $\int_a^b$ is reduced to the
above by the substitution x' = x - a.

2. Method

We approximate I by a sequence of approximate values
I_k, k = 0, 1, 2, ... The approximation I_k is obtained by dividing $[0, b]$ into 2^k congruent subintervals and evaluating the integral on each subinterval
by the midpoint formula. (The trapezoidal rule is
avoided because integrands frequently have singularities at the endpoints of the interval. Although harm-

less for the existence of the integral, these may
cause the program to fail; see example $\boxed{2}$.) Writing

$$h_k := 2^{-k} b \; ,$$

we have

$$I_k = h_k \sum_{m=0}^{2^k - 1} f((m + \tfrac{1}{2}) h_k) \; .$$

For any continuous f,

$$\lim_{k \to \infty} I_k = I \; .$$

If f is sufficiently smooth, the convergence of the
sequence $\{I_k\}$ may be sped up by the Romberg algorithm.
To this end the program saves the five currently most
recent values of I_k in locations in which they can be
used as input for the Romberg acceleration program
(see the following program) without external storage.

3. <u>Flow Diagram</u>

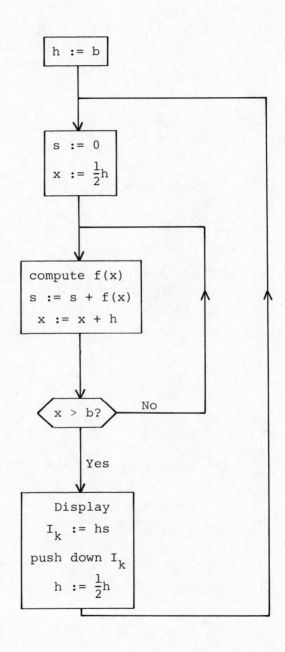

4. Storage and Program

R_0	R_1	R_2	R_3	R_4	R_5	R_6	R_7
I_0	I_1	I_2	I_3	I_4	h	x	b

	00		25	RCL 2
	01	RCL 7	26	STO 1
	02	STO 5	27	RCL 3
→	03	2	28	STO 2
	04	÷	29	RCL 4
	05	STO 6	30	STO 3
	06	CLX	31	RCL 5
	07	STO 4	32	GTO 03
	08	GTO 33	→ 33	
→	09	STO+4	34	
	10	RCL 5	35	
	11	STO+6	36	
	12	RCL 7	37	
	13	RCL 6	38	
	14	x < y	39	
	15	GTO 33	40	
	16	RCL 5	41	
	17	RCL 4	42	
	18	*	43	
	19	STO 4	44	
	20	PAUSE (R/S)	45	
	21	2	46	
	22	STO÷5	47	
	23	RCL 1	48	
	24	STO 0	49	

The program to compute f(x) should be in locations 33 through 49, assuming x in R_6. Only the stack may be used for temporary storage. The last instruction must be GTO 09. At this point, f(x) must be in the X register.

5. Operating Instructions

Load the main program. Load the program for computing f into locations 33 through 49; the last instruction must be GTO 09. Switch to RUN. Select a mode of displaying numbers, for instance

FIX 8

If f involves trigonometric functions with argument in radians, press

RAD

To start computation, press

PRGM
R/S

The calculator will pause briefly while displaying each I_k. If $k \overset{>}{=} 0$, and if

R/S

is pressed during pause, the values I_{k-m} will be found in R_{4-m}, m = 0, 1, 2, 3, 4.

6. Examples and Timing

1   b = 1, $f(x) := \dfrac{4}{1 + x^2}$. Program to compute f:

33	4
34	RCL 6
35	x^2
36	1
37	+
38	÷
39	GTO 09

The following values of I_k are obtained:

k	I_k		$I_4^{(k)}$
0	3.20000000		3.14191817
1	3.16235294		3.14159265
2	3.14680052	Romberg →	3.14159264
3	3.14289473		3.14159266
4	3.14191817		3.14159264

Time required 35 sec. Subjecting the values I_k to the Romberg algorithm produces the exact value π = 3.14159265 with an error of one unit in the last digit due to rounding effects.

2️⃣ $b = 1$, $f(x) := \dfrac{\text{Log}(1 + x)}{x}$. Program to compute f:

$$
\begin{array}{ll}
33 & 1 \\
34 & \text{RCL } 6 \\
35 & + \\
36 & \ln \\
37 & \text{RCL } 6 \\
38 & \div \\
39 & \text{GTO } 09
\end{array}
$$

Values obtained:

k	I_k		$I_4^{(k)}$
0	0.81093022		0.82241711
1	0.81936429		0.82246693
2	0.82167416	Romberg	0.82246702
3	0.82226766		0.82246702
4	0.82241711		0.82246702

Exact value: $\dfrac{\pi^2}{12} = 0.82246703$. Time required to generate $I_0, \ldots, I_4$ about 40 sec.

ROMBERG ALGORITHM

1. Purpose

To accelerate the convergence of a sequence $\{T_m\}$ to
its limit a_0, where

$$T_m := T(2^{-m}h) , \quad m = 0, 1, \ldots , 4 ,$$

and the function $T(h)$ has an asymptotic expansion of
the form

$$T(h) \approx a_0 + a_1 h^2 + a_2 h^4 + a_3 h^6 + \ldots , \quad h \to 0+ .$$

This problem poses itself in connection with numerical
integration $[T(h)$ = approximate value of integral calculated with step $h]$; in addition, it occurs in numerical differentiation, and in the computation of certain functions and constants. See examples below.

2. Method

The Romberg algorithm (see ENA, § 12.4; ACCA II,
§ 11.12). With the data

$$T_m := T(2^{-m}h) \ , \quad m = 0, 1, \ldots , 4$$

we generate a triangular array of numbers T_{mn} by the formulas

$$T_{m0} := T_m \ , \quad m = 0, 1, \ldots , 4 \ ;$$

and for $n = 1, 2, 3, 4$:

$$T_{mn} := \frac{4^n T_{m,n-1} - T_{m-1,n-1}}{4^n - 1} \ , \quad m = n, n+1, \ldots, 4. \quad (1)$$

If m were allowed to tend to infinity, one would have (loc. cit.) for each fixed n

$$T_{mn} = a_0 + O(4^{-mn}) \ .$$

Thus T_{44} may be regarded a much more accurate approximation to a_0 than T_{40} (mathematical error bounds are available). To check convergence, the program is arranged so that all values T_{4n} ($n = 0, 1, \ldots , 4$) are available at end of computation.

For reasons of numerical stability as well as economy, formula (1) is evaluated as

$$T_{mn} = T_{m,n-1} + \frac{1}{4^n - 1}(T_{m,n-1} - T_{m-1,n-1}) \ . \quad (2)$$

3. Flow Diagram

Because an address modification is not available on the HP-25, the operation (2) is always carried out at the same place. This is made possible by a cyclic permutation (here called "rotation") of the T_m.

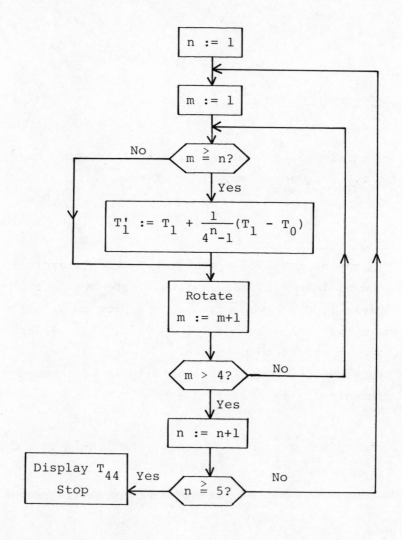

4. <u>Storage and Program</u>

R_0	R_1	R_2	R_3	R_4	R_5	R_6	R_7
T_0	T_1	T_2	T_3	T_4	temp	n	m

00		25	RCL 1
01	1	26	STO 0
02	STO 6	27	RCL 2
→ 03	1	28	STO 1
04	STO 7	29	RCL 3
→ 05	RCL 6	30	STO 2
06	RCL 7	31	RCL 4
07	x < y	32	STO 3
08	GTO 23	33	RCL 5
09	4	34	STO 4
10	RCL 6	35	1
11	y^x	36	STO+7
12	1	37	RCL 7
13	-	38	4
14	1/x	39	$x \stackrel{>}{=} y$
15	RCL 1	40	GTO 05
16	RCL 0	41	1
17	-	42	STO+6
18	*	43	RCL 6
19	RCL 1	44	5
20	+	45	$x \stackrel{>}{=} y$
21	STO 0	46	GTO 03
22	PAUSE (R/S)	47	RCL 0
→ 23	RCL 0	48	GTO 00
24	STO 5	49	

5. Operating Instructions

Load the program, and switch to RUN. Choose the desired mode of displaying numbers, for instance

SCI 8

to get floating eight-digit display. Load data $T_m = T(2^{-m}h)$ into R_m, m = 0, 1, ... , 4. (If the data are produced by the program "Numerical Integration with Step Refinement", they are already there.) When

PRGM
R/S

is pressed, the calculator will generate the array T_{mn} column by column, pausing briefly after computing each entry. (For an indeterminate display of each T_{mn}, change instruction 22 to R/S and press R/S after each display.) The calculator will stop by displaying T_{44}. T_{44} is also in R_0; generally, the elements in the last row of the array, $T_{4,n}$, are found in the registers R_{4-n}, n = 0, 1, ... , 4.

6. Examples and Timing

1 Numerical evaluation of

$$\int_0^1 x^9 \, dx$$

by the preceding program produces the approximate values

0.001953125
0.037544250
0.078839094
0.094288322
0.098544474

The present program accelerates the convergence of these values to produce the following values $T_{4,n}$ in the last row of the Romberg array (listed vertically here for reasons of space):

0.098544474
0.099963191
0.099998199
0.099999858
0.100000000

It is seen that the last value is accurate to all places given. Computing time about 44 sec.

2 Here we compute Euler's constant,

$$\gamma := \lim_{n \to \infty} (1 + \frac{1}{2} + \frac{1}{3} + \ldots + \frac{1}{n} - \text{Log } n) \ ,$$

directly from its definition, using Romberg acceleration. If for $n = 1, 2, \ldots$ we let

$$s(n) := 1 + \frac{1}{2} + \frac{1}{3} + \ldots + \frac{1}{n-1} + \frac{1}{2}\frac{1}{n} - \text{Log } n \, ,$$

then we evidently also have $\gamma = \lim s(n)$. The factor $\frac{1}{2}$ in front of $\frac{1}{n}$ is motivated by the fact that for suitable a_k

$$s(n) \approx \gamma + \sum_{k=1}^{\infty} a_k n^{-2k} \, , \qquad n \to \infty$$

(see ACCA II, § 11.11). Thus the numbers

$$g_k := s(2^k) \, , \qquad k = 0, 1, 2, \ldots \, ,$$

satisfy the conditions for applying the Romberg algorithm (see ACCA II, § 11.12). They are conveniently generated by the recurrence relation $g_0 := \frac{1}{2}$,

$$g_{k+1} = g_k + \sum_{2^k < m < 2^{k+1}} \frac{1}{m} + \frac{3}{2} 2^{-k-1} - \text{Log } 2 \, ,$$

$k = 0, 1, 2, \ldots$. It is an easy exercise to write a program that computes g_i for $i = 0, 1, \ldots, 4$ and stores it in R_i. We obtain

0.500000000
0.556852820
0.572038973
0.575915602
0.576890272

The program "Romberg Algorithm" accelerates these values as follows:

0.576890272
0.577215162
0.577215652
0.577215663
0.577215665

Again the last value is accurate to all places. Computing time for the acceleration about 44 sec.

3 For further examples on Romberg's algorithm see the programs "Numerical Integration with Step Refinement", "Plana Summation Formula", "Log-Arcsine Algorithm".

PLANA SUMMATION FORMULA

1. Purpose

To evaluate the integral

$$I := \int_0^\infty \frac{g(y)}{e^{2\pi y} - 1} \, dy$$

by means of numerical integration. This integral occurs, among other places, in the Plana summation formula (ACCA I, § 4.9),

$$\sum_{n=0}^\infty f(n) = \frac{1}{2}f(0) + \int_0^\infty f(x) \, dx + \int_0^\infty \frac{g(y)}{e^{2\pi y} - 1} \, dy \, ,$$

where

$$g(y) := - 2 \, \text{Im} \, f(iy) \, .$$

This formula holds, roughly, for functions f that are analytic for Re $z \gtreqless 0$ and grow less than $e^{2\pi|y|}$ for $y := \text{Im} \, z \to \pm\infty$. It may often be used to sum slowly converging series. The program works for any function g whose evaluation requires no more than 18 instructions.

2. Method

The integral I is approximated by the sums

$$I_m := \sum_{k=0}^{(\infty)} \frac{g(y_k)}{e^{2\pi y_k} - 1} \, , \quad m = 0, 1, 2, \ldots ,$$

where $h = h_0 \cdot 2^{-m}$, $y_k = \frac{h}{2} + kh$. The summation is stopped when $e^{2\pi y_k} - 1 > 1/\varepsilon$. The constants h_0 and $1/\varepsilon$ are input. The program displays each I_m after it has been computed. The convergence of the sequence $\{I_m\}$ may be accelerated by the Romberg algorithm (see the preceding algorithm).

3. <u>Flow Diagram</u>

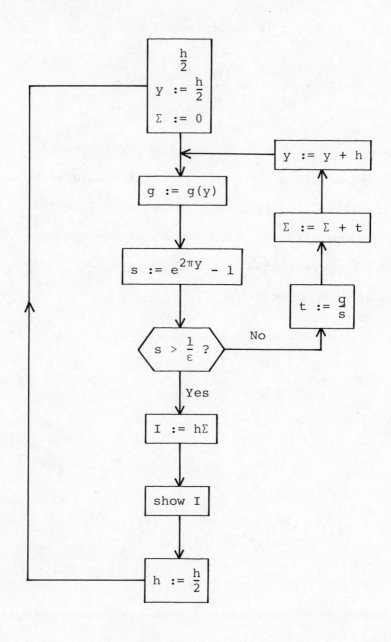

4. Storage and Program

R_0	R_1	R_2	R_3	R_4	R_5	R_6	R_7
h	$1/\varepsilon$	$\dfrac{h}{2}$			Σ	y	

	00		25	$1/x$
→	01	RCL 0	26	RCL 6
	02	2	27	2
	03	÷	28	π
	04	STO 2	29	*
	05	STO 6	30	*
	06	0	31	e^x
	07	STO 5	32	1
→	08	RCL 6	33	−
	09	3	34	RCL 1
	10	→ P	35	x < y
	11	ln	36	GTO 43
	12	→ P	37	R↓
	13	x^2	38	÷
	14	$x \gtrless y$	39	STO+5
	15	2	40	RCL 0
	16	*	41	STO+6
	17	RCL 6	42	GTO 08
	18	3	43	RCL 0
	19	→ P	44	RCL 5
	20	R↓	45	*
	21	+	46	R/S
	22	sin	47	RCL 2
	23	÷	48	STO 0
	24	*	49	GTO 01

Note: The function g is evaluated by instructions 08 through 25, presuming its argument y in R_6. g(y) should be in X register before executing instruction 26. In above program, g is taken from example 3 below.

5. Operating Instructions

After loading the program, switch to RUN. Indicate the mode of displaying numbers desired, for instance,

SCI 8

If function g involves trigonometric functions with arguments in radians (such as in examples 2 and 3 below), press

RAD

Load h_0 (such as 0.5) and press

STO 0

Load $1/\varepsilon$ (such as 10^{10}) and press

STO 1

When

PGRM

R/S

is pressed, the calculator will compute and display I_0. Pressing R/S again yields I_1, I_2, ...

The numbers I_0, I_1, ... are not stored. To subject them to the Romberg algorithm, they should be copied and used as input for the program Romberg Algorithm.

6. Examples and Timing

[1] Euler's constant γ (see preceding algorithm) may also be represented as

$$\gamma = \frac{1}{2} + \int\limits_0^\infty \frac{2y}{1 + y^2} \frac{1}{e^{2\pi y} - 1} \, dy$$

(ACCA I, p. 275). The function

$$g(y) := \frac{2y}{1 + y^2}$$

is generated by

08	RCL 6
09	2
10	*
11	RCL 6
12	x^2

13	1
14	+
15	÷
16	GTO 26

The following values I_m result by choosing $h_0 :=$
0.5, $1/\varepsilon := 10^{10}$:

$$6.6297262 * 10^{-2}$$
$$7.4582004 * 10^{-2}$$
$$7.6562828 * 10^{-2}$$
$$7.7052792 * 10^{-2}$$
$$7.7174967 * 10^{-2}$$

Total computing time: approx. 10 min. Subjecting
the foregoing values to the Romberg algorithm
yields the following last line of the Romberg
scheme:

7.7174967   7.7215692   7.7215663   7.7215664   7.7215664

(all $* 10^{-2}$). We conclude that $I = 7.7215664 *$
10^{-2}, and thus

$$\gamma = 0.577215664 ,$$

which is correct to within a rounding error of
$1 * 10^{-9}$.

2 Application of the Plana formula to $f(z) :=$ $(1 + z)^{-s}$ (principal value) shows that the Riemann zeta function (ACCA II, § 10.8),

$$\zeta(s) := \sum_{n=1}^{\infty} \frac{1}{n^s} ,$$

may be represented as follows:

$$\zeta(s) = \frac{1}{2} + \frac{1}{s - 1} + \int_0^{\infty} g(s; y) \frac{1}{e^{2\pi y} - 1} dy ,$$

where

$$g(s; y) := \frac{2 \sin(s \text{ Arctg } y)}{(1 + y^2)^{s/2}} .$$

This representation holds for all $s \neq 1$. We load s into R_3 and use the following program to compute g:

08	RCL 6
09	$\tan^{-1}$
10	RCL 3
11	*
12	sin
13	2
14	*
15	RCL 6
16	x^2

$$
\begin{array}{ll}
17 & 1 \\
18 & + \\
19 & \text{RCL } 3 \\
20 & y^x \\
21 & \sqrt{x} \\
22 & \div \\
23 & \text{GTO } 26
\end{array}
$$

For $s := 1.1$, $h_0 := 0.5$, $1/\varepsilon := 10^{10}$ we obtain

$$
\left.
\begin{array}{l}
7.2387260 * 10^{-2} \\
8.1548623 * 10^{-2} \\
8.3730175 * 10^{-2} \\
8.4269295 * 10^{-2} \\
8.4403697 * 10^{-2}
\end{array}
\right\}
\xrightarrow{\text{Romberg}}
\left\{
\begin{array}{l}
8.4403697 * 10^{-2} \\
8.4448497 * 10^{-2} \\
8.4448464 * 10^{-2} \\
8.4448464 * 10^{-2} \\
8.4448464 * 10^{-2}
\end{array}
\right.
$$

There follows $\zeta(1.1) = 10.584448464$. The program may be tested by evaluating $\zeta(s)$ at the "trivial" zeros $s = -2, -4, \ldots$.

3 To evaluate the very slowly converging series

$$
S := \sum_{n=3}^{\infty} \frac{1}{n\{\text{Log } n\}^2} ,
$$

we use the Plana formula where

$$
f(z) := \frac{1}{(3 + z)\{\text{Log}(3 + z)\}^2} .
$$

This requires the evaluation of

$$g(y) := - 2 \ \text{Im} \ f(iy) \ ,$$

which may be written as

$$g(y) = \frac{2 \ \sin(\alpha + 2\beta)}{rR^2} \ ,$$

where

$$3 + iy = r \ e^{i\alpha} \ , \ \text{Log} \ r + i\alpha = R \ e^{i\beta} \ .$$

The program is given in the foregoing main program; it offers interesting applications of the → P operation. The final multiplication by 2 is omitted for lack of space. The following values result with $h_0 := 0.5$, $1/\varepsilon := 10^{10}$:

$$
\left.
\begin{array}{l}
0.8983867 * 10^{-2} \\
1.0025014 * 10^{-2} \\
1.0280286 * 10^{-2} \\
1.0343783 * 10^{-2} \\
1.0359638 * 10^{-2}
\end{array}
\right\}
\xrightarrow{\text{Romberg}}
\left\{
\begin{array}{l}
1.0359638 * 10^{-2} \\
1.0364923 * 10^{-2} \\
1.0364921 * 10^{-2} \\
1.0364921 * 10^{-2} \\
1.0364921 * 10^{-2}
\end{array}
\right.
$$

We conclude

$$S = \frac{1}{2} \ \frac{1}{3(\text{Log} \ 3)^2} + \frac{1}{\text{Log} \ 3} + 2 * 1.0364921 * 10^{-2}$$

$$= 1.0690583 \ .$$

The total time required for this computation is
about 20 min.

DIFFERENTIAL EQUATION OF FIRST ORDER:
TRAPEZOIDAL METHOD

1. Purpose

To integrate numerically a general first order equation

$$y' = f(x,y) \tag{1}$$

subject to the initial condition

$$y(x_0) = y_0 \ .$$

2. Method

The trapezoidal rule. Choosing an integration step h, we determine approximate values y_n of the values $y(x_n)$ of the exact solution $y(x)$ at the points $x_n := x_0 + nh$ (n = 1, 2, ...) by the formula

$$y_{n+1} - y_n = \frac{h}{2}\left[f(x_n,y_n) + f(x_{n+1},y_{n+1})\right] \ . \tag{2}$$

The implicit equation for y_{n+1} that arises at each

step is solved by iteration, using Euler's method as a predictor (see ENA, § 14.7). Iteration is stopped when two successive iterates differ by less than ε. Eleven storage locations are available to evaluate the expression $\frac{h}{2}f(x,y)$.

The order of the method defined by (2) is 2, independently of the number of iterations performed. That is, as $h \to 0$, the approximate values y_n at a fixed value of x_n tend to the exact value $y(x_n)$ with an error of $O(h^2)$. If ε is small (say, $\varepsilon = 10^{-10}$) and the equation (2) thus is satisfied to full machine accuracy, the convergence of y_n to $y(x_n)$ may be sped up by the Romberg algorithm (see the example). However, a large number of iterations may then be necessary at each step. The number of iterations may be reduced by giving ε larger values. For very large values of ε (say, $\varepsilon = 10^{10}$), the corrector will be applied only once, and the method becomes identical with the simplified Runge-Kutta method,

$$y_{n+1} - y_n = \frac{h}{2}\left[f(x_n,y_n) + f(x_{n+1},y_n+hf(x_n,y_n))\right]. \quad (3)$$

The order of the method is still 2, but Romberg's h^2-extrapolation is no longer feasible.

3. Flow Diagram

We write $x := x_n$, $y := y_n$; y^* denotes the current argument y in f(x,y), y^+ is the corrected value.

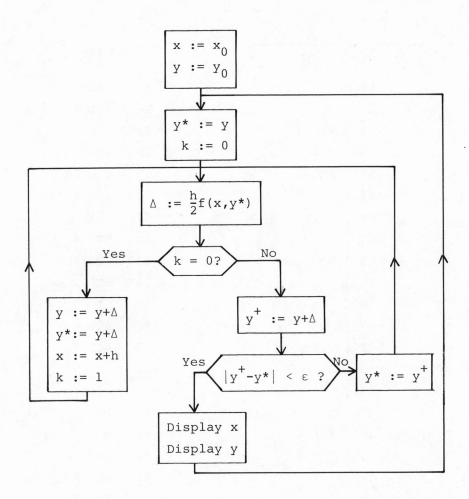

A "flag" k is used to distinguish between the predictor (k = 0) and the corrector cycles (k = 1).

4. Storage and Program

R_0	R_1	R_2	R_3	R_4	R_5	R_6	R_7
h	x_n	y_n	ε	k		f(y)	y

00		25	STO+2
01	CLX	26	RCL 2
02	STO 4	27	+
03	RCL 2	28	STO 7
04	STO 7	29	RCL 0
→ 05		30	STO+1
06		31	1
07		32	STO 4
08	Program to	33	GTO 05
09		→ 34	RCL 2
10	compute	35	STO+6
11		36	RCL 6
12	f(x,y*)	37	RCL 7
13		38	−
14		39	ABS
15		40	RCL 3
16	RCL 0	41	x $\overset{>}{=}$ y
17	*	42	GTO 46
18	2	43	RCL 6
19	÷	44	STO 7
20	STO 6	45	GTO 05
21	RCL 4	→ 46	RCL 1
22	x ≠ 0	47	PAUSE
23	GTO 34	48	RCL 6
24	RCL 6	49	STO 2

5. Operating Instructions

Load the main program. Load the program for computing
$f(x,y^*)$ into program locations $05 \div 15$, assuming x in
R_1 and y^* in R_7. This program must place $f(x,y^*)$ into
X. (As an alternate possibility, one may use locations
$05 \div 20$ for a program that places $\Delta := \frac{h}{2}f(x,y^*)$ into
R_6.) R_5 is available to store a constant required for
the evaluation of f.

   Move the operating switch to RUN. Select the desired mode of presenting results, for instance by
pressing SCI 6 to get floating six-digit display. Load

integration step h	into	R_0
starting value x_0	into	R_1
starting value y_0	into	R_2
tolerance ε for stopping inner iteration	into	R_3

When

PRGM

R/S

is pressed, the calculator will start computing. After
completing inner iteration, it will briefly display x_1
and stop by displaying y_1. Pressing

R/S

will cause the display of x_2, y_2, etc. The process may be repeated until the desired range of x_n has been covered.

To check the accuracy of the results, the computation should be repeated with the steps $\frac{h}{2}$, $\frac{h}{4}$, ... , until the results have converged to the desired accuracy. If ε is small, the convergence may be sped up by the Romberg algorithm.

6. Example and Timing

The exact equation of the mathematical pendulum of length ℓ,

$$y" = \frac{g}{\ell} \sin y$$

(g := gravitational constant = 9.81 ms^{-2}) may under the initial condition $y'(0) = 0$ be integrated to give

$$y' = -\sqrt{\frac{2g}{\ell}} \sqrt{\cos y - \cos y_0} ,$$

where $y_0 = y(0)$. We solve this equation under the initial condition

$$y_0 = y(0) = \frac{\pi}{3} = 1.047198 .$$

Here

$$\Delta = \frac{h}{2}f(x,y) = -h\sqrt{\frac{g}{2\ell}}\sqrt{\cos y - 0.5} \ .$$

The following program is used to compute Δ:

05	RCL 7
06	cos
07	.
08	5
09	-
10	$\sqrt{x}$
11	RCL 5
12	*
13	RCL 0
14	*
15	CHS
16	STO 6
17	NOP
18	NOP
19	NOP
20	NOP

The constant $\sqrt{\frac{g}{2\ell}}$ is assumed in R_5. Since the argument of cos is assumed in radians, RAD must be pressed before starting computation.

The following values result if $\ell = 1$, beginning with $h = 0.1$ and successively halving the step:

| | | | Y | |
x	h=0.1	h=0.05	h=0.025	h=0.0125
0.0	1.047198	1.047198	1.047198	1.047198
0.1	1.005241	1.004979	1.004913	1.004896
0.2	0.881649	0.880558	0.880284	0.880215
0.3	0.683836	0.681290	0.680648	0.680488
0.4	0.425723	0.421160	0.420008	0.419719
0.5	0.128475	0.121689	0.119976	0.119546
0.6	-0.180863	-0.189433	-0.191595	-0.192137

Applying Romberg to the values at x = 0.5 yields

0.128475			
0.121689	0.119427		
0.119976	0.119405	0.119404	
0.119546	0.119402	0.119402	0.119402

For h = 0.1 and an iteration tolerance of $\varepsilon = 10^{-8}$, the time per integration step varies between 15 and 60 sec, due to the considerable number of iterations required. By setting $\varepsilon = 10^{+8}$, the program is modified to correct only once, possibly at the expense of stability. The time per integration step is then reduced to 6 sec.

AUTONOMOUS DIFFERENTIAL EQUATION OF SECOND ORDER,
FIRST DERIVATIVE ABSENT

1. Purpose

To integrate numerically a differential equation of
the special form

$$y'' = f(y) \ , \tag{1}$$

subject to the initial conditions

$$y(x_0) = y_0 \ , \qquad y'(x_0) = z_0 \ .$$

Equations of the special form (1) occur, among other
places, in problems of non-linear vibrations without
damping.

2. Method

A simplified version of the Runge-Kutta method (see
ENA, § 14.5). Let h be the integration step, and let
$x_n := x_0 + nh$. Approximate values y_n of $y(x_n)$ and z_n
of $y'(x_n)$ are computed by the formulas

$$y_{n+1} = y_n + hz_n + \frac{h^2}{2}f(y_n) ,$$

(2)

$$z_{n+1} = z_n + hf(y_n + \frac{h}{2}z_n) ,$$

$n = 0, 1, 2, \ldots$. The error in y_n and z_n as $h \to 0$ and $x_n = x$ is fixed is $O(h^2)$.

3. Flow Diagram

We write $x := x_n$, $y := y_n$, $z := z_n$; y^* denotes the current argument in f; $\Delta := hf(y^*)$.

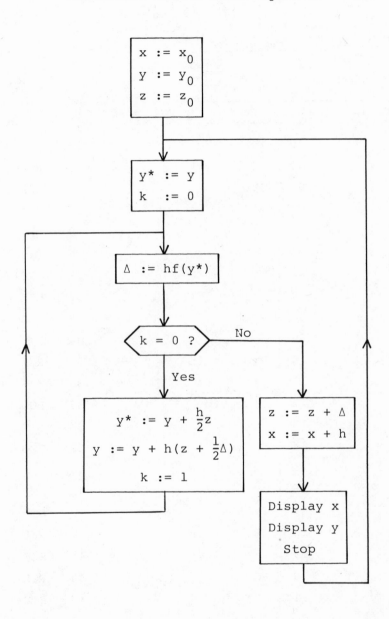

4. Storage and Program

R_0	R_1	R_2	R_3	R_4	R_5	R_6	R_7
h	x	y	z	k		Δ	y*

00		25	GTO 43
01	CLX	26	RCL 3
02	STO 4	27	RCL 0
03	RCL 2	28	*
04	STO 7	29	2
→ 05 ⎫		30	÷
06 ⎪		31	STO+7
07 ⎪		32	RCL 6
08 ⎪		33	2
09 ⎪		34	÷
10 ⎪	Program to	35	RCL 3
11 ⎪		36	+
12 ⎬	compute	37	RCL 0
13 ⎪		38	*
14 ⎪	f(y*)	39	STO+2
15 ⎪		40	1
16 ⎪		41	STO 4
17 ⎪		42	GTO 05
18 ⎪		→ 43	RCL 6
19 ⎭		44	STO+3
20	RCL 0	45	RCL 0
21	*	46	STO+1
22	STO 6	47	RCL 1
23	RCL 4	48	PAUSE
24	x ≠ 0	49	RCL 2

5. Operating Instructions

Load the main program. Load the program for calculating $f(y*)$ into program locations 05 ÷ 19, assuming $y*$ in R_7. Program must put $f(y*)$ into X. R_5 is available to store a constant needed for the calculation of f.

Move the operating switch to RUN. Select a mode of displaying numbers. If f contains trigonometric functions with arguments expressed in radians, press

RAD

Load data as follows:

h	into	R_0
x_0	into	R_1
y_0	into	R_2
z_0	into	R_3

Put auxiliary constants into R_5. Pressing

PRGM

R/S

will start calculation. The calculator will briefly display x_1 and stop by displaying y_1. To exhibit z_1, press

RCL 3

To continue computation, press

$$R/S$$

This will cause display of x_2 and y_2, etc.

6. Example and Timing

We integrate the exact equation of the mathematical pendulum,

$$y" = - \frac{g}{\ell} \sin y ,$$

where $g/\ell = 9.81$, under the initial conditions

$$y(0) = \frac{\pi}{3} , \quad y'(0) = 0$$

(compare the program "Trapezoidal Rule"). The results for various values of h are as follows:

x	h=0.1	h=0.05	h=0.025	h=0.0125
		y		
0.0	1.047198	1.047198	1.047198	1.047198
0.1	1.004719	1.004785	1.004859	1.004885
0.2	0.878363	0.879600	0.880033	0.880156
0.3	0.675164	0.678906	0.680034	0.680340
0.4	0.408987	0.416708	0.418874	0.419443
0.5	0.102068	0.114850	0.118251	0.119124
0.6	-0.216063	-0.198319	-0.193810	-0.192679

Applying h^2-extrapolation to the values at $x = 0.5$
yields

0.102068	
0.114850	0.119110
0.118251	0.119384
0.119124	0.119415

The theoretical hypotheses for continuing Romberg ex-
trapolation beyond the second column of the Romberg
scheme are not satisfied here.

Computing time per integration step (including
display of x) approximately 5 sec, independently of
choice of h.

LINEAR SECOND ORDER DIFFERENTIAL EQUATION

1. Purpose

To integrate numerically the general linear second order equation,

$$y'' + p(x)y' + q(x)y = 0 , \qquad (1)$$

subject to initial conditions

$$y(x_0) = y_0 , \qquad y'(x_0) = y_0' . \qquad (2)$$

2. Method

Finite differences. Denoting by y_n an approximate value of the exact solution $y(x)$ at the point $x_n := x_0 + nh$, where h is the integration step, we approximate

$$y'(x_n) \qquad \text{by} \qquad \frac{y_{n+1} - y_{n-1}}{2h}$$

and

$$y''(x_n) \qquad \text{by} \qquad \frac{y_{n+1} - 2y_n + y_{n-1}}{h^2} .$$

Introducing these approximations in the differential equation and writing

$$p_n := p(x_n) \ , \quad q_n := q(x_n) \ ,$$

there results

$$\frac{1}{h^2}(y_{n+1} - 2y_n + y_{n-1}) + \frac{p_n}{2h}(y_{n+1} - y_{n-1}) + q_n y_n = 0 \ ,$$

which can be solved for y_{n+1} to yield

$$y_{n+1} = \frac{1}{1 + \frac{h}{2}p_n}\{(2 - h^2 q_n)y_n - (1 - \frac{h}{2}p_n)y_{n-1}\}$$

$$\tag{3}$$

$$= \frac{1}{2 + hp_n}\{(4 - 2h^2 q_n)y_n - (2 - hp_n)y_{n-1}\} \ .$$

This is used to calculate the approximations y_n recursively. The starting value $y_0 = y(x_0)$ is given. To obtain y_1, we use Taylor's expansion in the form

$$y_1 = y_0 + hy_0' + \frac{h^2}{2}y_0'' \ .$$

Here y_0 and $y_0' = y'(x_0)$ are given, and y_0'' can be calculated from the differential equation. There results

$$y_1 = y_0 + hy_0' - \frac{h^2}{2}[p_0 y_0' + q_0 y_0] \ . \tag{4}$$

This calculation must be performed manually.

3. Flow Diagram

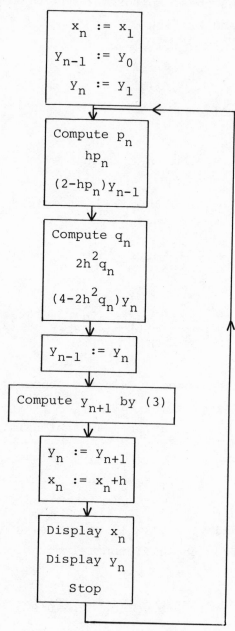

4. Storage and Program

R_0	R_1	R_2	R_3	R_4	R_5	R_6	R_7
h	x_1	y_0	y_1	temp	temp		
	x_n	y_{n-1}	y_n				

00		25	
01	⎫	26	RCL 0
02	⎪	27	x^2
03	Program to	28	*
04	compute	29	2
05	⎬	30	*
06	$p(x_n)$	31	CHS
07	⎪	32	4
08	⎭	33	+
09	RCL 0	34	RCL 3
10	*	35	STO 2
11	STO 4	36	*
12	2	37	RCL 5
13	RCL 4	38	−
14	−	39	RCL 4
15	RCL 2	40	2
16	*	41	+
17	STO 5	42	÷
18	⎫	43	RCL 0
19	⎪	44	RCL 1
20	Program to	45	+
21	compute	46	STO 1
22	⎬	47	PAUSE
23	$q(x_n)$	48	R ↓
24	⎪	49	STO 3

5. Operating Instructions

Load the program, including the programs to compute
$p(x)$ (locations 01 through 08) and $q(x)$ (locations 18
through 25), assuming x in R_1. The registers R_6 and R_7
are available to store constants needed in the calculation of p and q.

Move the operating switch to RUN. Select the mode
of displaying numbers, and choose correct mode for arguments of trigonometric functions (ordinarily by
pressing RAD) if such functions occur in p or q. Load
data as follows:

$$h \qquad \text{into} \quad R_0$$
$$x_1 \ (:=x_0+h) \quad \text{into} \quad R_1$$
$$y_0 \qquad \text{into} \quad R_2$$

Calculate y_1 from (4), and load it into R_3. Pressing

PRGM

R/S

starts computation. The calculator briefly displays x_2
and stops while displaying y_2. Computation is continued by pressing

R/S

which results in display of x_3, y_3, etc. Continue in

this fashion until the desired range of x values has been covered.

As always with differential equation problems, it is a good practice to repeat calculation with $h := \frac{1}{2}h$ in order to check convergence of values y_n at fixed values of x. Romberg acceleration may be used at least once.

6. Example and Timing

We obtain values of the Bessel function $J_1(x)$ by integrating the differential equation satisfied by it,

$$y'' + \frac{1}{x} y' + (1 - \frac{1}{x^2}) y = 0 ,$$

using

$$y_0 = J_1(0) = 0 , \quad y_0' = J_1'(0) = \frac{1}{2} .$$

No harm is done by the singularities of the coefficient functions p and q at x = 0, because the program uses these functions only from $x = x_1$ onward. Equation (4), however, cannot be used to find y_1. From the power series defining $J_1(x)$ we get to the same order of accuracy

$$y_1 = \frac{h}{2} .$$

The programs to calculate p and q are

01	RCL 1
02	1/x

and

18	1
19	RCL 1
20	x^2
21	1/x
22	-

The remaining locations are to be filled up with NOP instructions. (As an alternate possibility, the whole program could be compressed, which is not problematic here because there are no internal references. In this case, the instruction GTO 00 should be inserted at the end.)

Values obtained using several different steps h are as follows:

	y			$J_1(x)$
x	h=0.1	h=0.05	h=0.025	(exact values)
0.0	0.000000	0.000000	0.000000	0.000000
0.1	0.050000	0.049958	0.049944	0.049938
0.2	0.099667	0.099553	0.099517	0.099501
0.3	0.148603	0.148406	0.148345	0.148319
0.4	0.196436	0.196150	0.196063	0.196027

0.5	0.242806	0.242429	0.242315	0.242268
0.6	0.287368	0.286899	0.286758	0.286701
0.7	0.329790	0.329230	0.329063	0.328996
	. . . .		
3.6	0.095396	0.095459	0.095466	0.095466
3.7	0.053631	0.053789	0.053824	0.053834
3.8	0.012490	0.012740	0.012801	0.012821
3.9	-0.027698	-0.027360	-0.027274	-0.027244
4.0	-0.066613	-0.066193	-0.066082	-0.066043

Computing time per integration step approx. 4 sec.

Part 6

SPECIAL FUNCTIONS

LOG-ARCSINE ALGORITHM

1. Purpose

To compute the functions Log x and Arcsin x for arbitrary x > 0 and for arbitrary x ε $[-1, 1]$, respectively, by an algorithm requiring only the evaluation of square roots.

2. Method

Numerical differentiation (see ACCA II, § 11.12). Clearly, for x > 0,

$$\text{Log } x = \frac{\partial}{\partial h} \left. x^h \right|_{h=0} .$$

Therefore, if

$$\ell(h) := \frac{1}{2h}(x^h - x^{-h})$$

and

$$\ell_n := \ell(2^{-n}) , \quad n = 0, 1, 2, \ldots ,$$

then

$$\lim_{n\to\infty} \ell_n = \text{Log } x \ .$$

Similarly, if $x \in [-1, 1]$ and $\alpha := \text{Arcsin } x$, then

$$\text{Arcsin } x = \frac{\partial}{\partial h} \sin(h\alpha) \Big|_{h=0} \ .$$

Thus by putting

$$a(h) := \frac{\sin(h\alpha)}{h}$$

and

$$a_n := a(2^{-n}) \ , \quad n = 0, 1, 2, \ldots \ ,$$

we have

$$\lim_{n\to\infty} a_n = \text{Arcsin } x \ .$$

The sequences $s_n = \ell_n$ and $s_n = a_n$ both satisfy the same recurrence relation

$$s_{n+1} = s_n \sqrt{\frac{2s_n}{s_n + s_{n-1}}} \ , \quad n = 0, 1, \ldots \ ; \qquad (1)$$

if

$$s_{-1} := \frac{1}{4}(x^2 - \frac{1}{x^2}) \,, \quad s_0 := \frac{1}{2}(x - \frac{1}{x}) \,, \qquad (2)$$

then $s_n = \ell_n$, and if

$$s_{-1} := x\sqrt{1 - x^2} \,, \quad s_0 := x \,, \qquad (3)$$

then $s_n = a_n$. In both cases, the convergence of the sequence $\{s_n\}$ can be sped up by the Romberg algorithm. The foregoing algorithm fails for the trivial values Log 1 = 0 and Arcsin 0 = 0, but it is very stable otherwise.

3. Flow Diagram

The program computes $\{\ell_n\}$ if the index k is set equal to 0, and it computes $\{a_n\}$ if $k \neq 0$. In either case it saves the five most recent elements of the sequence $\{s_n\}$.

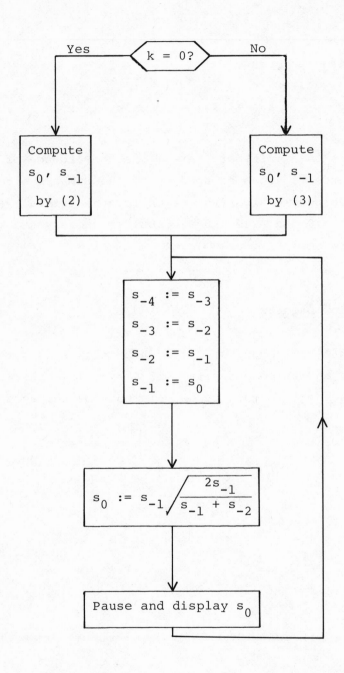

4. Storage and Program

R_0	R_1	R_2	R_3	R_4	R_5	R_6	R_7
s_{-4}	s_{-3}	s_{-2}	s_{-1}	s_0		x	$\boxed{k}$

00			25	1/x
01	STO 6		26	-
02	RCL 7		27	4
03	x = 0		28	÷
04	GTO 15		29	STO 3
05	1	→	30	RCL 1
06	RCL 6		31	STO 0
07	STO 4		32	RCL 2
08	x^2		33	STO 1
09	-		34	RCL 3
10	$\sqrt{x}$		35	STO 2
11	RCL 6		36	RCL 4
12	*		37	STO 3
13	STO 3		38	2
14	GTO 30		39	*
→ 15	RCL 6		40	RCL 3
16	ENTER		41	RCL 2
17	1/x		42	+
18	-		43	÷
19	2		44	$\sqrt{x}$
20	÷		45	RCL 3
21	STO 4		46	*
22	RCL 6		47	STO 4
23	x^2		48	PAUSE
24	ENTER		49	GTO 30

5. Operating Instructions

Load the program, and switch to RUN. Select the mode of displaying numbers, such as

SCI 8

to get the floating eight-digit display. Load k into R_7, x into X ($x \neq 1$ if $k = 0$ and $x \neq 0$ if $k \neq 0$). Pressing

PRGM
R/S

will start computation. The calculator pauses briefly while displaying each s_n. No convergence test is provided; convergence must be determined by inspection. Press

R/S

during pause to recover five most recent values. These will be in R_0 through R_4, the correct locations for the Romberg algorithm. The most recent value is also displayed in the X register.

6. Examples and Timing

1̄ k := 0, x = 10 yields with FIX 9 display

Log 10 = 2.302585094   in 16 iterations.

Time required about 30 sec. There is an error of
one unit in the last digit. By using SCI 8 display we get

Log 100 = 4.6051702   in 13 iterations,
Log 1000 = 6.9077552   in 16 iterations.

2 k := 0, x := e yields the elements s_{-1}, s_0, ... ,
s_3 of the sequence $\{s_n\}$ as follows:

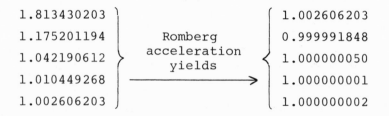

1.813430203		1.002606203
1.175201194	Romberg	0.999991848
1.042190612	acceleration yields	1.000000050
1.010449268	$\longrightarrow$	1.000000001
1.002606203		1.000000002

Clearly, a FIX 8 display would have given the
correct answer Log e = 1.00000000.

3 k = 1, x := 1 yields with FIX 9 display

Arcsin 1 = 1.570796319   in 15 iterations,

which differs from the exact value $\pi/2$ =
1.570796327 by $8 * 10^{-9}$.

THE GAMMA FUNCTION

1. Purpose

To compute the values of the gamma function,

$$\Gamma(x) := \lim_{n \to \infty} \frac{n^x n!}{(x)_{n+1}}$$

$$= \int_0^\infty e^{-t} t^{x-1} \, dt \quad (x > 0)$$

for arbitrary real values of x with a relative error $< 10^{-8}$.

2. Method

Stirling's formula (see ACCA II, § 8.5). We select n such that Stirling's formula evaluates $\Gamma(x+n)$ with the required low relative error and then obtain

$$\Gamma(x) = \frac{\Gamma(x+n)}{(x)_n}$$

(see ACCA II, § 8.4). Stirling's formula is usually

given in the form

$$\Gamma(x) = \sqrt{2\pi}\ e^{(x - 1/2)\text{Log } x - x + J(x)}\ , \qquad (1)$$

where $J(x)$ is the Binet function,

$$J(x) := \frac{1}{\pi} \int_0^\infty \frac{x}{t^2 + x^2} \text{Log } \frac{1}{1 - e^{-2\pi t}}\ dt\ ,$$

$x > 0$. It is known that $J(x)$ possesses an enveloping asymptotic expansion as $x \to \infty$,

$$J(x) \approx \frac{1}{12\ x} - \frac{1}{360\ x^3} + \frac{1}{1260\ x^5} - \cdots\ . \qquad (2)$$

The direct implementation of Stirling's formula, approximating $J(x)$ by the terms of the asymptotic series shown above, would produce the required accuracy for $x+n > 5$, but the resulting program would be too long for the HP-25. We therefore modify the formula, as follows. Writing (1) in the form

$$\Gamma(x) = \sqrt{\frac{2\pi}{x}}\ e^{x(\text{Log } x - f(x))}\ , \qquad (3)$$

we see that it is necessary to evaluate the function

$$f(x) := 1 - \frac{J(x)}{x}\ , \qquad (4)$$

which within the prescribed tolerance may be approximated by the polynomial in $1/x^2$,

$$g(x) := 1 - \frac{1}{12\ x^2} + \frac{1}{360\ x^4} - \frac{1}{1260\ x^6} \qquad (5)$$

(see also Table 1).

To evaluate $g(x)$ directly by means of (5), using any of the available methods for computing the values of a polynomial, would require the storage of three constants with a total of 9 digits, occupying 9 program memory locations. The resulting program would again exceed the capacity of the HP-25.

It turns out that a shorter program is obtained by converting the function $g(x)$ into a continued fraction (i.e., by representing it by its diagonal Padé approximant), using the qd algorithm (see ACCA, § 12.4). The following qd table results from the four coefficients of the function g:

a_n	$q_1^{(n)}$	$e_1^{(n)}$	$q_2^{(n)}$
1			
	$-\dfrac{1}{12}$		
$-\dfrac{1}{12}$		$\dfrac{1}{20}$	
	$-\dfrac{1}{30}$		$\dfrac{53}{315}$
$\dfrac{1}{360}$		$-\dfrac{53}{210}$	
	$-\dfrac{2}{7}$		
$-\dfrac{1}{1260}$			

This yields the continued fraction

$$\cfrac{x}{x - \cfrac{-\dfrac{1}{12}}{x - \cfrac{\dfrac{1}{20}}{x - \cfrac{\dfrac{53}{315}}{x}}}}\,,$$

which agrees with $g(x)$ up to an error of at most $O(x^{-8})$ as $x \to \infty$. In the above fraction we may replace the unwieldy $\frac{53}{315}$ by $\frac{53}{318} = \frac{1}{6}$ without affecting the accuracy of the final result. ($\frac{1}{6}$ would arise exactly if the coefficient $\frac{1}{1260}$ in $g(x)$ were replaced by $\frac{17}{21600} \sim \frac{1}{1270}$.) By algebraic manipulation the resulting approximation to $f(x)$,

$$h(x) := \cfrac{x}{x + \cfrac{\dfrac{1}{12}}{x - \cfrac{\dfrac{1}{20}}{x - \cfrac{\dfrac{1}{6}}{x}}}}\,,$$

may be simplified as follows:

$$h(x) = \cfrac{1}{1 + \cfrac{1}{3x\left(4x + \cfrac{1}{5\left(\dfrac{1}{6x} - x\right)}\right)}}\,.$$

This, finally, is the approximation to f(x) used in
our program. Only 4 one-digit constants need to be
stored. The numerical values given in Table 1 show
that the error committed in approximating g(x) by
h(x) for x $\overset{>}{=}$ 5 again lies in the permitted range. If
x < 5, our program thus will determine the integer n
mentioned initially such that 5 $\overset{<}{=}$ x+n < 6.

x	f(x)	g(x)	h(x)
1	0.918938533	0.918650794	0.918566776
2	0.979329652	0.979327877	0.979327701
3	0.990774025	0.990773946	0.990773914
4	0.994802332	0.994802324	0.994802324
5	0.996671061	0.996671060	0.996671061
6	0.997687312	0.997687312	0.997687312
7	0.998300470	0.998300470	0.998300470
8	0.998698592	0.998698592	0.998698592

Table 1

3. Flow Diagram

We let y := x+n, p := $(x)_n$.

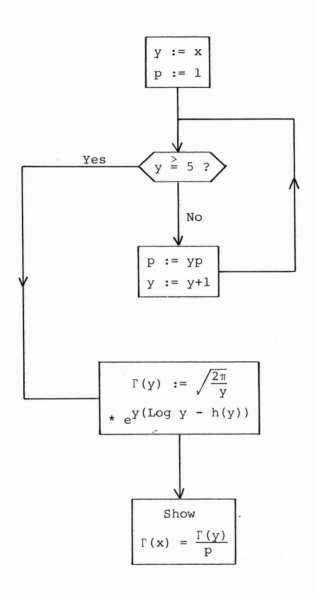

4. Storage and Program

R_0	R_1	R_2	R_3	R_4	R_5	R_6	R_7
x	p	y					

00		25	RCL 2
01	STO 0	26	*
02	1	27	3
03	STO 1	28	*
04	5	29	1/x
05	RCL 0	30	1
→ 06	x $\overset{>}{=}$ y	31	+
07	GTO 12	32	1/x
08	STO*1	33	CHS
09	1	34	RCL 2
10	+	35	ln
11	GTO 06	36	+
→ 12	STO 2	37	RCL 2
13	6	38	*
14	*	39	e^x
15	1/x	40	2
16	RCL 2	41	π
17	-	42	*
18	5	43	RCL 2
19	*	44	÷
20	1/x	45	$\sqrt{x}$
21	RCL 2	46	*
22	4	47	RCL 1
23	*	48	÷
24	+	49	GTO 00

5. Operating Instructions

Load the program; operating switch to RUN. Select a
mode of displaying numbers, ordinarily by pressing

SCI 8

to get 8 digit floating display. Load x into X register. Pressing

PRGM

R/S

will cause the value of $\Gamma(x)$ to be displayed. Values
are generally correct to 8 significant digits; for
large x errors may be slightly larger due to inaccu-
racies in exponential function.

To get $\Gamma(x)$ for new value of x, load new x into X.
Pressing

R/S

will yield new value.

6. Examples and Timing

1	x = 1 yields $\Gamma(x)$ = 1.0000000

1   x = 1   yields $\Gamma(x)$ = 1.0000000

$\qquad\qquad$ 2 $\qquad\qquad\qquad\qquad\qquad\quad$ 1.0000000

$\qquad\qquad$ 3 $\qquad\qquad\qquad\qquad\qquad\quad$ 2.0000000

$\qquad\qquad$ 4 $\qquad\qquad\qquad\qquad\qquad\quad$ 6.0000000

$\qquad\qquad$ 5 $\qquad\qquad\qquad\qquad\qquad\quad$ $2.4000000 * 10^1$

$\qquad\qquad$ 6 $\qquad\qquad\qquad\qquad\qquad\quad$ $1.2000000 * 10^2$

$\qquad\qquad$ 7 $\qquad\qquad\qquad\qquad\qquad\quad$ $7.1999999 * 10^2$

$\qquad\qquad$ 8 $\qquad\qquad\qquad\qquad\qquad\quad$ $5.0400000 * 10^3$

$\qquad\qquad$ 9 $\qquad\qquad\qquad\qquad\qquad\quad$ $4.0319999 * 10^4$

$\qquad\quad$ 10 $\qquad\qquad\qquad\qquad\qquad\quad$ $3.6287999 * 10^5$

These values provide a genuine check, because the relation $\Gamma(n+1)$ = n! is not used in the program. Computing time for each value about 5 sec.

2   x = 0.5 yields $\Gamma(\frac{1}{2})$ = 1.7724538, which agrees with the exact value, $\sqrt{\pi}$.

3   x = 0 correctly yields "Error", because $\Gamma(x)$ has a pole at x = 0. The same holds for all negative integers.

4   x = -2.37 yields $\Gamma(-2.37)$ = 1.2183595.

INCOMPLETE GAMMA FUNCTION

1. Purpose

To compute, for arbitrary $x > 0$ and arbitrary real a, the values of the incomplete gamma function,

$$\Gamma(a,x) := \int_x^\infty t^{a-1} e^{-t}\, dt . \qquad (1)$$

2. Method

Expansion in terms of reciprocal Laguerre polynomials. Let

$$L_n^{(a)}(x) := \sum_{k=0}^{n} \binom{n+a}{n-k} \frac{(-x)^k}{k!}$$

denote the Laguerre polynomial of order n with parameter a (see ACCA I, p. 119). Then there holds for all $x > 0$ and all $a < 1$

$$\Gamma(a,x) = x^a e^{-x} \sum_{n=0}^{\infty} \frac{(1-a)_n}{(n+1)!} \frac{1}{L_n^{(-a)}(-x)\, L_{n+1}^{(-a)}(-x)} \qquad (2)$$

(see ACCA II, § 12.13). This expansion also holds for
a $\overset{>}{=}$ 1, provided -x is not a zero of one of the polynomials $L_n^{(-a)}$; see example $\boxed{4}$ for how to deal with the
exceptional case. Using the asymptotic theory of the
Laguerre polynomials (see Szegö, Orthogonal polynomials, Theorem 8.22.5) the n-th term of the above series
(including the factor $x^a e^{-x}$) can be shown to be asymptotic to

$$4\pi \sqrt{\frac{x}{n}}\, e^{-4\sqrt{nx}} \ .$$

The convergence of the series (2) thus is subgeometric.
More precisely, the number n of terms required to get
k correct decimal digits is asymptotically equal to

$$n = 0.33\,\frac{k^2}{x} \tag{3}$$

and thus is roughly proportional to k^2, independent of
a, and inversely proportional to x.

    For purposes of computation we write (2) in the
form

$$\Gamma(a,x) = \sum_{n=1}^{\infty} \frac{\gamma_n}{\ell_{n-1}\ell_n} \ , \tag{4}$$

where

$$\gamma_n := x^a e^{-x}\,\frac{(1-a)_{n-1}}{(2)_{n-1}}$$

and

$$\ell_n := L_n^{(-a)}(-x) \ .$$

The quantities γ_n are computed by the recurrence

$$\gamma_1 := x^a e^{-x} \ , \quad \gamma_{n+1} := \gamma_n \frac{n-a}{n+1} \ . \tag{5}$$

From the recurrence relation of the Laguerre polynomials (see Szegö, Orthogonal polynomials, equation (5.1.10)) we find

$$\ell_{-1} = 0 \ , \quad \ell_0 = 1 \ ,$$

$$\ell_n = \frac{1}{n}\{(2n - a - 1 + x)\ell_{n-1} - (n - a - 1)\ell_{n-2}\} \ ,$$

$n = 1, 2, \ldots$. To save arithmetical operations, this is used in the form

$$\ell_n = \frac{1}{n}\{(n - a - 1)(\ell_{n-1} - \ell_{n-2}) + (n + x)\ell_{n-1}\} . \tag{6}$$

Summation is stopped if

$$\left| \frac{\gamma_n}{\ell_n \ell_{n-1}} \right| < \varepsilon \ ,$$

where ε may be chosen arbitrarily.

3. <u>Flow Diagram</u>

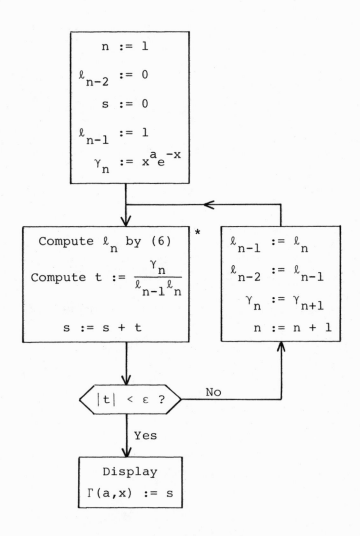

To save operations, most operations indicated in the
n := n + 1 box are anticipated in the box marked *.

4. Storage and Program

R_0	R_1	R_2	R_3	R_4	R_5	R_6	R_7
a	x	ℓ_{n-2}	ℓ_{n-1}	s	γ_n	n	ε

00			25	RCL 6
01	CLX		26	RCL 1
02	STO 2		27	+
03	STO 4		28	RCL 3
04	1		29	STO 2
05	STO 3		30	*
06	STO 6		31	+
07	RCL 1		32	RCL 6
08	RCL 0		33	÷
09	y^x		34	STO 3
10	STO 5		35	RCL 2
11	RCL 1		36	*
12	e^x		37	÷
13	STO÷5		38	STO+4
→ 14	RCL 5		39	PAUSE
15	RCL 6		40	ABS
16	RCL 0		41	RCL 7
17	−		42	$x \overset{>}{=} y$
18	STO*5		43	GTO 49
19	1		44	1
20	−		45	STO+6
21	RCL 3		46	RCL 6
22	RCL 2		47	STO÷5
23	−		48	GTO 14
24	*		→ 49	RCL 4

5. Operating Instructions

Load the program, move the operating switch to RUN.
Select a mode of displaying results, for instance by
pressing

FIX

9

Load data as follows:

$$a \quad \text{into} \quad R_0$$
$$x \quad \text{into} \quad R_1$$
$$\varepsilon \quad \text{into} \quad R_7$$

For 9 digit accuracy, let $\varepsilon = 10^{-10}$. Pressing

PRGM

R/S

will start computation. The calculator briefly displays each term of the series (4) and stops if that
term is $< \varepsilon$, displaying value of $\Gamma(a,x)$. To get new
value, it suffices to load new data and to press

R/S

6. Examples and Timing

The incomplete gamma function is of interest because
it contains as special cases a number of important
functions of mathematical physics and of statistics.
The following special cases occur in the Handbook of
Mathematical Functions by Abramowitz and Stegun.

$\boxed{1}$ The exponential integral,

$$E_1(x) := \int_x^\infty \frac{e^{-t}}{t} \, dt \ ,$$

obviously is given by

$$E_1(x) = \Gamma(0,x) \ .$$

Using the tolerance $\varepsilon = 10^{-10}$, the following values are obtained for different values of x:

x	$E_1(x)$	number n of terms required	computing time in sec
0.5	0.559773594	67	190
1.0	0.219383934	36	100
2.0	0.048900511	19	54
4.0	0.003779352	10	27

The computed values are in error by at most one
unit in the last place.

2   The <u>exponential integrals of higher order</u>,

$$E_n(x) := \int_1^\infty \frac{e^{-xt}}{t^n} \, dt \, , \quad n = 1, 2, \ldots \, ,$$

may be expressed as

$$E_n(x) = x^{n-1} \, \Gamma(1-n, x) \, .$$

Some sample values computed by our program (using $\varepsilon = 10^{-9}$) are as follows:

x	$E_2(x)$	$E_4(x)$	$E_{20}(x)$
0.5	0.3266439	0.1652428	0.0310612
1.0	0.1484955	0.0860625	0.0183460
2.0	0.0375343	0.0250228	0.0064143
5.0	0.0009965	0.0007830	0.0002783

All values are correct to the number of digits given. Computing times are comparable to those given under 1 .

3   The <u>error function</u>,

$$\text{erf}(x) = \frac{2}{\sqrt{\pi}} \int_0^x e^{-t^2} \, dt \, ,$$

equals

$$\text{erf}(x) = 1 - \frac{1}{\sqrt{\pi}} \, \Gamma(\tfrac{1}{2}, x^2) \, ,$$

as may be seen by a simple substitution. Using the tolerance $\varepsilon = 10^{-10}$, the program permits the following values to be obtained:

x	erf(x)	number n of terms used	computing time in sec
1.0	0.842700793	34	97
2.0	0.995322265	10	30
3.0	0.999977910	5	15

The values given are correct; however, the program Error function is much more efficient for $x < 2$.

4 Chi-square probability function. This function may be defined as

$$Q(\chi^2 | \nu) := \frac{\Gamma(\frac{\nu}{2}, \frac{\chi^2}{2})}{\Gamma(\frac{\nu}{2})} .$$

Some values provided by our program (using $\varepsilon = 10^{-6}$) are as follows:

ν	χ^2			
	1	2	3	4
2	0.60653	0.36788	0.22313	0.13534
4	0.90980	0.73578*	0.55783	0.40601
6	0.98561	0.91970	0.80885	0.67669*
8	0.99825	0.98101	0.93436	0.85712

As mentioned in section 2, for each $a \gtreqless 1$ the program may fail for isolated values of x due to negative zeros of the Laguerre polynomials. For instance, in the above table $(\chi^2 | \nu) = (2|4)$ is such a pair of values. Failure to compute $\Gamma(a,x)$ at $x = x_0$ is recognized by an error halt. In such cases, $\Gamma(a,x_0)$ may be obtained as the arithmetic mean of the values $\Gamma(a,x_0 \pm \delta)$ where δ is suffi- ciently small. In the above table, the values marked * were obtained in this manner. Of course, near failure may occur for values near x_0. This is recognized by the appearance of large terms in the series (4). Again, more precise values can then be obtained by linear interpolation.

ERROR FUNCTION

1. Purpose

To compute the error function,

$$\operatorname{erf}(x) := \frac{2}{\sqrt{\pi}} \int_0^x e^{-t^2} dt ,\qquad (1)$$

for all $x \geq 0$ with an error of less than 10^{-9}.

2. Method

A power series is used for $0 \leq x < 2$, and a continued fraction for $x \geq 2$.

    a) The Taylor series $x = 0$,

$$\operatorname{erf}(x) = \frac{2}{\sqrt{\pi}} \sum_{n=0}^{\infty} \frac{(-1)^n}{n!(2n+1)} x^{2n+1} ,\qquad (2)$$

is not to be recommended for numerical computation, be-cause the large alternating terms may cause cancella-tion, and because the general term is awkward to com-pute recursively. However, the series (2) may be expressed as a confluent hypergeometric series (ACCA I, § 1.5),

$$\text{erf}(x) = \frac{2}{\sqrt{\pi}} \ x \ {}_1F_1 \ (\tfrac{1}{2} \ ; \ \tfrac{3}{2} \ ; \ -x^2) \ .$$

We thus may use Kummer's first identity (ACCA I, eq. (1.5-3); ACCA II, eq. (9.7-8)) to obtain

$$\text{erf}(x) = \frac{2x}{\sqrt{\pi}} \ e^{-x^2} \ {}_1F_1 \ (1 \ ; \ \tfrac{3}{2} \ ; \ x^2) \ . \qquad (3)$$

In the last series, all terms are positive, and no cancellation can occur. Moreover, if

$$\frac{2x}{\sqrt{\pi}} \ e^{-x^2} \ {}_1F_1 \ (1 \ ; \ \tfrac{3}{2} \ ; \ x^2) = \sum_{n=0}^{\infty} a_n \ ,$$

then

$$a_0 = \frac{2x}{\sqrt{\pi}} \ e^{-x^2} \ ,$$

$$a_n = \frac{x^2}{n + \tfrac{1}{2}} \ a_{n-1} \ ,$$

$n = 1, 2, \ldots$. This formula is used in the program to evaluate the series (3) for $0 \overset{<}{=} x < 2$.

b) Numerical experiments show that the series (3) can be used safely to compute $\text{erf}(x)$ at least up to x = 5.0, where erf(x) = 1 to the accuracy required. However, the time required to compute erf(x) by (3) is about 10x seconds for $x \stackrel{>}{=} 2$ due to the increasing number of terms that have to be summed. For $x \stackrel{>}{=} 2$ we therefore use an alternate method that is considerably faster. For all x > 0 there holds

$$erf(x) = 1 - erfc(x) \ ,$$

where

$$erfc(x) := \frac{2}{\sqrt{\pi}} \int_x^\infty e^{-t^2} dt$$

is the complementary error function. This function admits the asymptotic expansion

$$erfc(x) \approx \frac{e^{-x^2}}{\sqrt{\pi} \ x} \ _2F_0(\frac{1}{2} \ , \ 1 \ ; \ - \frac{1}{x^2}) \ , \ x \to \infty \ , \qquad (4)$$

which however cannot be used to obtain arbitrarily accurate values for any x. An expression that converges for all x > 0 is obtained by converting the asymptotic series into a continued fraction (ACCA II, § 12.13). This yields

$$erfc(x) = \frac{e^{-x^2}}{\sqrt{\pi}} \ \{\frac{1}{|x} + \frac{\frac{1}{2}}{|x} + \frac{1}{|x} + \frac{\frac{3}{2}}{|x} + \frac{2}{|x} + \ldots \} \ . \qquad (5)$$

It has been determined experimentally that for $x \stackrel{>}{=} 2$

the 16th approximant yields erfc(x) with an error < 10^{-9}. Thus our program uses the approximation

$$\text{erf}(x) = 1 - \frac{e^{-x^2}}{\sqrt{\pi}} \{\frac{1}{x} + \frac{\frac{1}{2}}{x} + \ldots + \frac{8}{x}\} .$$

Working with a fixed number of approximants has the advantage that the continued fraction can be evaluated rapidly working from the tail upward (ACCA II, § 12.1).

3. Flow Diagram

We set $a := a_n$, $s := \Sigma a_n$, $v := n + \frac{1}{2}$, $m := 8 - \frac{1}{2}k$.

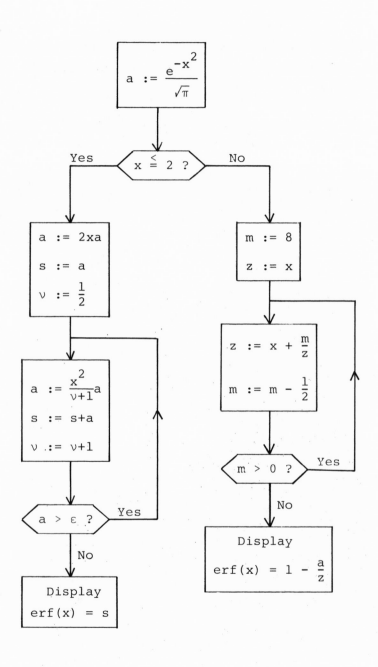

4. Storage and Program

R_0	R_1	R_2	R_3	R_4	R_5	R_6	R_7
$\boxed{x}$	a	ν	s	m	$\sqrt{\pi}$	.5	ε

00		25	STO*1
01	STO 0	26	RCL 1
02	x^2	27	STO+3
03	CHS	28	RCL 7
04	e^x	29	x < y
05	RCL 5	30	GTO 18
06	÷	31	RCL 3
07	STO 1	32	GTO 00
08	2	→ 33	RCL 0
09	RCL 0	34	8
10	$x \overset{>}{=} y$	→ 35	STO 4
11	GTO 33	36	$x \overset{<}{>} y$
12	STO*1	37	÷
13	RCL 6	38	RCL 0
14	STO 2	39	+
15	STO÷1	40	RCL 4
16	RCL 1	41	RCL 6
17	STO 3	42	−
→ 18	RCL 0	43	x ≠ 0
19	x^2	44	GTO 35
20	RCL 2	45	R ↓
21	1	46	STO÷1
22	+	47	1
23	STO 2	48	RCL 1
24	÷	49	−

5. Operating Instructions

Load the program; move the operating switch to RUN.
Press

FIX

9

for nine-digit display (results will be accurate to 9
digits). Load $\varepsilon := 10^{-10}$ into R_7:

EEX

CHS

1

0

STO 7

Load 0.5 into R_6:

.

5

STO 6

Load $\sqrt{\pi}$ into R_5:

π

$\sqrt{x}$

STO 5

To obtain erf(x) for arbitrary $x \geq 0$, load x into X register and press

PRGM

R/S

Value of erf(x) will be displayed after a few seconds. To obtain erf(x) for new x, load new x into X and press

R/S

6. Examples and Timing

1   x = 1.99 yields erf(x) = 0.995111413, time 20 sec.
2   x = 2.00 yields erf(x) = 0.995322265, time 11 sec.
3   x = 0.01 yields erf(x) = 0.011283416, time 3 sec.
4   x = 5.00 yields erf(x) = 1.000000000, time 11 sec.
5   The changeover point from power series to continued fraction may be changed to any integer k such that 0 < k < 10, by substituting k for 2 in instruction 08. It is thus possible to compare the performance of the formulas (3) and (5) for various values of x. For instance, for x = 4 we get

by (3)   erf(x) = 0.999999985 , time 40 sec,
by (5)   erf(x) = 0.999999985 , time 11 sec.

COMPLETE ELLIPTIC INTEGRALS

1. Purpose

To compute, for $0 \leqq k < 1$, the complete elliptic integral of the first kind,

$$K(k) := \int_0^{\pi/2} \frac{1}{\sqrt{1 - k^2 \sin^2 \phi}} \, d\phi \ ,$$

and of the second kind,

$$E(k) := \int_0^{\pi/2} \sqrt{1 - k^2 \sin^2 \phi} \, d\phi \ .$$

2. Method

The arithmetic-geometric mean. Let α_0, β_0 be positive numbers, and let

$$\alpha_{n+1} := \frac{1}{2}(\alpha_n + \beta_n) \ , \quad \beta_{n+1} := \sqrt{\alpha_n \beta_n} \ ,$$

$n = 0, 1, 2, \ldots$. Then the sequences $\{\alpha_n\}$ and $\{\beta_n\}$

rapidly converge to a common limit $\mu(\alpha_0,\beta_0)$, called the arithmetic-geometric mean of α_0 and β_0. If $\alpha_0 = 1$, $\beta_0 = k' := \sqrt{1 - k^2}$, then

$$K(k) = \frac{\pi}{2\mu(1,k')} .$$

Moreover, if $\gamma_0 := k$,

$$\gamma_{n+1} := \frac{1}{2}(\alpha_n - \beta_n) ,$$

$n = 0, 1, 2, \ldots$, then

$$E(k) = K(k)\{1 - \frac{1}{2} \sum_{m=0}^{\infty} 2^m \gamma_m^2\}$$

(see ACCA II, § 9.10, problems 9 - 13).

3. Flow Diagram

We let $\alpha := \alpha_n$, $\beta := \beta_n$, $\alpha' := \alpha_{n+1}$, $\gamma := \gamma_n$, $\sigma := \sum 2^k \gamma_k^2$, $\tau := 2^n$.

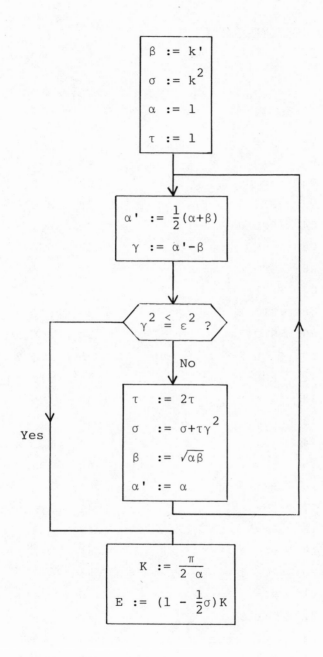

4. Storage and Program

R_0	R_1	R_2	R_3	R_4	R_5	R_6	R_7
k	ε^2	α_n	β_n	γ_n^2	α_{n+1}	σ_n	2^n
k in X							

00			25	RCL 7
01	$\sin^{-1}$		26	RCL 4
02	cos		27	$*$
03	STO 3		28	STO+6
04	RCL 0		29	RCL 2
05	x^2		30	RCL 3
06	STO 6		31	$*$
07	1		32	$\sqrt{x}$
08	STO 2		33	STO 3
09	STO 7		34	RCL 5
→ 10	RCL 2		35	STO 2
11	RCL 3		36	GTO 10
12	+		→ 37	π
13	2		38	2
14	÷		39	RCL 2
15	STO 5		40	$*$
16	RCL 3		41	÷
17	−		42	STO 5
18	x^2		43	1
19	STO 4		44	RCL 6
20	RCL 1		45	2
21	$x \overset{\geq}{=} y$		46	÷
22	GTO 37		47	−
23	2		48	$*$
24	STO$*$7		49	RCL 5

5. Operating Instructions

Load the program; move the operating switch to RUN.
Select a mode of displaying numbers, for instance by
pressing

FIX 9

to get 9 digits after decimal point. Load $\varepsilon^2 = 10^{-19}$
into R_1, as follows:

EEX
CHS
1
9
STO 1

Load k into R_0 and also leave it in X register. Pressing

PRGM
R/S

will after a few seconds cause K(k) to be displayed.
To display E(k), press

$$x \overset{>}{\underset{<}{}} y$$

K(k) then will be in Y register.

6. Examples and Timing

For $k = 1/\sqrt{2} = 0.707106781$ we get

$$K = 1.854074677 \; , \quad E = 1.350643881 \; ;$$

computing time approx. 8 sec. For $k = 0.9999$ the results are

$$K = 5.645148167 \; , \quad E = 1.000514502 \; ;$$

computing time approx. 12 sec.

BESSEL FUNCTIONS, INTEGER ORDER

1. Purpose

To compute the Bessel functions $J_n(x)$ for arbitrary $x > 0$ and for the orders $n = 0, 1, 2, 3, 4$.

2. Method

J.C.P. Miller's method of backward recurrence [see W. Gautschi, Computational aspects of three-term recurrence relations, SIAM Review 9, 24 - 82 (1967)]. We use a recurrence relation satisfied by the sequence $\{J_n(x)\}$ (x fixed) in the form

$$J_{n-1} = \frac{2n}{x} J_n - J_{n+1} \tag{1}$$

If the recurrence

$$Y_{n-1} = \frac{2n}{x} Y_n - Y_{n+1} \tag{2}$$

is solved with arbitrary starting values, say,

$$y_{m+1} := 0 , \quad y_m := 1 ,$$

where the integer m is sufficiently large, then the elements of the sequence y_{m-n}, y_{m-2}, ... soon become proportional to J_{m-1}, J_{m-2}, ... , because $\{J_n\}$ is a "recessive" solution of the difference equation (2). Thus, approximately,

$$J_n = \frac{1}{c} y_n , \tag{3}$$

where c is independent of n. The factor c may be determined from the known relation

$$J_0(x) + 2J_2(x) + 2J_4(x) + \ldots = 1 ,$$

which gives

$$c = y_0 + 2y_2 + 2y_4 + \ldots . \tag{4}$$

In this program m may be selected as an arbitrary positive integer; see the examples for a rule of thumb on how to select m as a function of x. The program then constructs the sequence $\{y_n\}$, always saving the last five elements. The values $J_n(x)$ are then computed from (3), approximating c by

$$y_0 + 2y_2 + 2y_4 + \ldots + 2y_{2\lceil m/2 \rceil} .$$

3. <u>Flow Diagram</u>

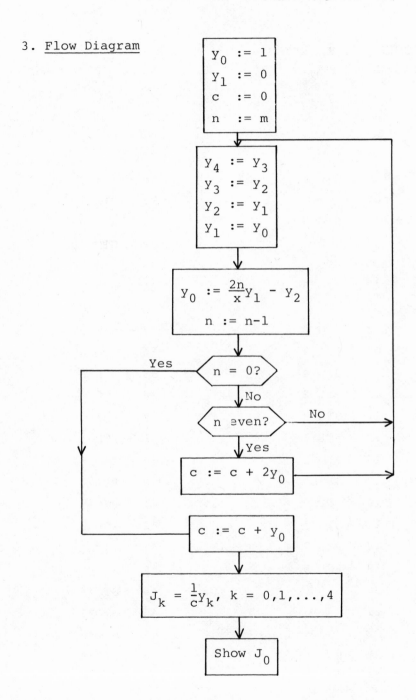

4. Storage and Program

R_0	R_1	R_2	R_3	R_4	R_5	R_6	R_7
J_0	J_1	J_2	J_3	J_4	c	x	$\boxed{m}$
							2n

00			25	2
01	STO 6		26	–
02	2		27	STO 7
03	STO*7		28	x = 0
04	0		29	GTO 40
05	STO 1		30	4
06	STO 5		31	÷
07	1		32	FRAC
08	STO 0		33	x ≠ 0
→ 09	RCL 3		34	GTO 09
10	STO 4		35	RCL 0
11	RCL 2		36	2
12	STO 3		37	*
13	RCL 1		38	STO+5
14	STO 2		39	GTO 09
15	CHS		→ 40	RCL 0
16	RCL 0		41	STO+5
17	STO 1		42	RCL 5
18	RCL 7		43	1/x
19	RCL 6		44	STO*0
20	÷		45	STO*1
21	*		46	STO*2
22	+		47	STO*3
23	STO 0		48	STO*4
24	RCL 7		49	RCL 0

5. Operating Instructions

Load the program; turn the operating switch to RUN.
Select the mode of displaying numbers, for example,
by pressing

$$SCI\ 8$$

Load m into R_7. (m = 30 is adequate for values x < 10.)
Load x into X. Press

$$PRGM$$
$$R/S$$

to start the computation. The calculator will stop by
displaying calculated value of $J_0(x)$; the values $J_k(x)$
(k = 0, 1, ... , 4) are also found in R_k. To restart
the computation with a new value of x, simply reload
m into R_7, the new x into X, and press

$$R/S$$

6. Examples and Timing

[1]   For x = 1, m = 30 we get the values

$$J_0 = 0.76519769$$
$$J_1 = 0.44005059$$

$$J_2 = 0.11490348$$
$$J_3 = 0.01956335$$
$$J_4 = 0.00247664$$

which are exact to the number of digits given. Computing time 40 sec.

2 For x = 10, m = 30 the results are

$$J_0 = -0.24593576$$
$$J_1 = 0.04347275$$
$$J_2 = 0.25463031$$
$$J_3 = 0.05837938$$
$$J_4 = -0.21960269$$

correct to all digits given.

3 For x = 17.5 we choose m = 40. The computing time now is about 59 sec. There results

$$J_0 = -0.10311040$$
$$J_1 = -0.16341997$$
$$J_2 = 0.08443383$$
$$J_3 = 0.18271913$$
$$J_4 = -0.02178727$$

Again, the errors are $< 10^{-8}$. The value m = 50, which theoretically produces still more accurate results, causes overflow.

BESSEL FUNCTIONS, ARBITRARY ORDER

1. Purpose

To compute, with an error $< 10^{-8}$, the values of the Bessel function $J_\nu(x)$ of arbitrary real order ν for $0 \leq x \leq 10$.

2. Method

Power series expansion. We have (see ACCA II, § 9.7)

$$J_\nu(x) = \frac{(\frac{x}{2})^\nu}{\Gamma(\nu+1)} \sum_{n=0}^{\infty} \frac{(-1)^n (\frac{x^2}{4})^n}{(\nu+1)_n n!} \ .$$

We write this in the form $a \sum b_n$, where

$$a := \frac{(\frac{x}{2})^\nu}{\Gamma(\nu+1)} \ , \quad b_0 := 1 \ ,$$

and the remaining b_n are calculated recursively by

$$b_n := - \frac{(\frac{x}{2})^2}{n(\nu+n)} b_{n-1} \ , \quad n = 1, 2, \ldots \ .$$

The summation is terminated if $|b_n| < 10^{-12}$. The term $\Gamma(\nu+1)$ required in the computation of a must be computed separately, if necessary by the program gamma function.

3. Flow Diagram

We write $s := \Sigma\ b_n$.

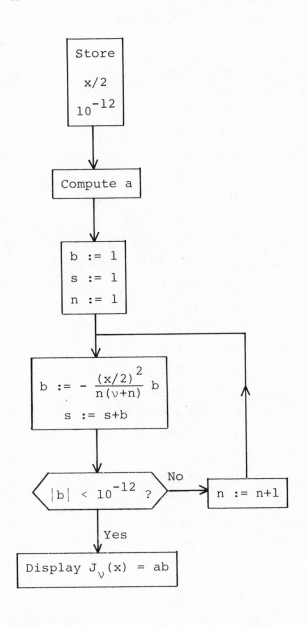

4. Storage and Program

R_0	R_1	R_2	R_3	R_4	R_5	R_6	R_7
ν	$\boxed{\Gamma(\nu+1)}$	10^{-12}	a	Σ	b	n	$\frac{x}{2}$

00		25	RCL 6
01	2	26	*
02	÷	27	÷
03	STO 7	28	CHS
04	1	29	STO*5
05	EEX	30	RCL 5
06	CHS	31	STO+4
07	1	32	ABS
08	2	33	RCL 2
09	STO 2	34	x $\overset{\geq}{=}$ y
10	RCL 7	35	GTO 39
11	RCL 0	36	1
12	y^x	37	STO+6
13	RCL 1	38	GTO 20
14	÷	→ 39	RCL 3
15	STO 3	40	RCL 4
16	1	41	*
17	STO 4	42	GTO 00
18	STO 5	43	
19	STO 6	44	
→ 20	RCL 7	45	
21	x^2	46	
22	RCL 6	47	
23	RCL 0	48	
24	+	49	

5. Operating Instructions

Load the program, and move the switch to RUN. Press

$$FIX\ 8$$

to get a reliable eight-digit output. Load

$$\nu \quad\quad \text{into} \quad R_0$$
$$\Gamma(\nu+1) \quad \text{into} \quad R_1 \ .$$

(The latter value may be computed by the program gamma function.) Load x into the X register. When

$$PRGM$$
$$R/S$$

is pressed, the calculation will start. The calculator stops by displaying $J_\nu(x)$.

This program does not work if ν is a negative integer, $\nu = -n$. Here the relation

$$J_{-n}(x) = (-1)^n J_n(x)$$

should be used. The program may be inaccurate if ν is close to a negative integer.

For nonintegral ν, the Bessel function of the second kind may be computed using the relation

$$Y_{\nu}(x) = \frac{J_{\nu}(x) \cos\nu\pi - J_{-\nu}(x)}{\sin\nu\pi} .$$

For integer values of ν, $\nu = n$, the limit of the fore-going as $\nu \to n$ may be calculated numerically, using the Romberg algorithm (see example $\boxed{4}$).

6. Examples and Timing

$\boxed{1}$ $J_0(2) = 0.22389078$, all digits correct, time required about 13 sec.

$\boxed{2}$ $J_0(10) = -0.24593576$, all digits correct, time required about 27 sec.

$\boxed{3}$ To compute $J_{3/2}(5)$, we require

$$\Gamma(\tfrac{5}{2}) = \tfrac{1}{2} \cdot \tfrac{3}{2} \sqrt{\pi} .$$

Computation yields $J_{3/2}(5) = -0.16965131$, in agreement with the exact value.

$\boxed{4}$ To compute $Y_0(x)$, we may use the relation

$$Y_0(x) = \frac{2}{\pi} \frac{\partial}{\partial \nu} J_{\nu}(x) \Big|_{\nu=0} .$$

Letting

$$d_{\nu} := \frac{J_{\nu}(x) - J_{-\nu}(x)}{2\nu} ,$$

our program yields for x = 2

ν	d_ν
2	0.00000000
1	0.57672481
0.5	0.74780184
0.25	0.78844828
0.125	0.79839964

With the Romberg algorithm the limit of these values as $\nu \to 0$ is determined as 0.80169622. Thus
we obtain

$$Y_0(2) = 0.51037566 \; ,$$

which is in error by only $1 * 10^{-8}$.

BESSEL FUNCTIONS: ASYMPTOTIC SERIES

1. Purpose

To compute the asymptotic series, for arbitrary $x > 0$, of the Bessel functions of both the first and second kind,

$$J_\nu(x) \quad \text{and} \quad Y_\nu(x) ,$$

for arbitrary real orders ν.

2. Method

The asymptotic series is computed most conveniently from the formula

$$J_\nu(x) + i\, Y_\nu(x) \sim \sqrt{\frac{2}{\pi x}}\, e^{i(x - \frac{\nu\pi}{2} - \frac{\pi}{4})} \,\, {}_2F_0(\tfrac{1}{2} + \nu, \tfrac{1}{2} - \nu;\, \tfrac{1}{2ix}), \quad (1)$$

where ${}_2F_0$ is a hypergeometric series (see ACCA II, § 11.5). We write

$$ {}_2F_0 = \Sigma\, (-i)^n a_n = R + iS ,$$

and calculate the a_n by means of the recurrence

$$a_0 := 1, \quad a_n := \frac{(-\frac{1}{2} + \nu + n)(-\frac{1}{2} - \nu + n)}{2xn} a_{n-1} .$$

The sums R and S are then given by

$$R = a_0 - a_2 + a_4 - a_6 + \cdots ,$$

$$S = - a_1 + a_3 - a_5 + a_7 - \cdots .$$

The point of termination of these series may be chosen arbitrarily. Ordinarily, to get best results, one should terminate as soon as $|a_n| > |a_{n-1}|$. If ν equals an integer plus $1\!/2$, the sums R and S will terminate mathematically, and the representation given by (1) is not only asymptotic but exact.

To save programming storage, the correct assignment of the terms $\pm a_n$ to the sums R and S is left to the operator (see operating instructions). It then becomes possible to compress the whole program into 49 steps by converting R + iS into polar coordinates,

$$R + iS = T e^{i\phi} \quad (T > 0) ,$$

which yields

$$e^{i(x - \frac{\nu\pi}{2} - \frac{\pi}{4})} (R + iS) = T e^{i(x - \frac{\nu\pi}{2} - \frac{\pi}{4} + \phi)} .$$

Converting back to cartesian coordinates then yields

$$J_\nu(x) \stackrel{\sim}{\sim} \sqrt{\frac{2}{\pi x}} \, T \cos(x - \frac{\nu\pi}{2} - \frac{\pi}{4} + \phi) \, , \qquad (2)$$

$$Y_\nu(x) \stackrel{\sim}{\sim} \sqrt{\frac{2}{\pi x}} \, T \sin(x - \frac{\nu\pi}{2} - \frac{\pi}{4} + \phi) \, . \qquad (3)$$

3. Flow Diagram

We set $q := (-\frac{1}{2}+\nu+n)(-\frac{1}{2}-\nu+n)/2xn$.

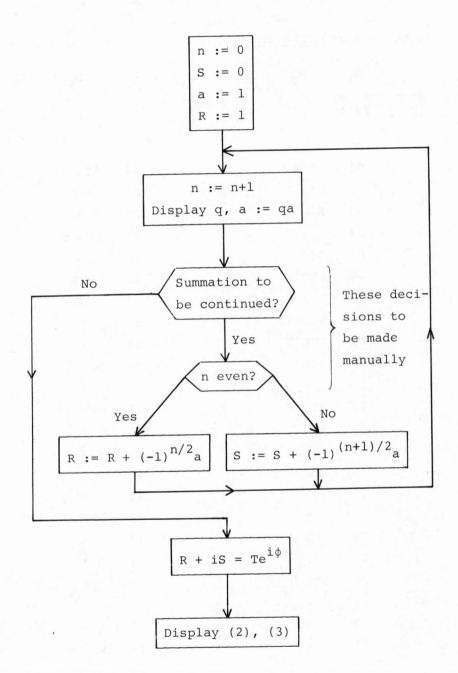

4. Storage and Program

R_0	R_1	R_2	R_3	R_4	R_5	R_6	R_7
x	$2\nu+1$	2n	a_n	R	S		

00		25	STO*3
01	CLX	26	RCL 3
02	STO 2	27	R/S
03	STO 5	28	GTO 07
04	1	→ 29	RCL 5
05	STO 3	30	RCL 4
06	STO 4	31	→ P
→ 07	2	32	2
08	STO+2	33	π
09	RCL 2	34	÷
10	RCL 1	35	RCL 0
11	−	36	÷
12	RCL 2	37	$\sqrt{x}$
13	RCL 1	38	*
14	+	39	x $\gtrless$ y
15	2	40	RCL 0
16	−	41	+
17	*	42	RCL 1
18	RCL 2	43	π
19	÷	44	*
20	RCL 0	45	4
21	4	46	÷
22	*	47	−
23	÷	48	x $\gtrless$ y
24	PAUSE	49	→ R

5. Operating Instructions

Load the program, and switch to RUN. Press

$$FIX \ 8$$

to obtain adequate display of results. Press

$$RAD$$

to cause arguments of trigonometric functions to be
interpreted in radians. Load

$$x \quad \text{into} \quad R_0$$
$$2\nu+1 \quad \text{into} \quad R_1$$

(the index ν must be stored in this form for reasons
of space). Pressing

$$PRGM$$
$$R/S$$

will first cause q_n to be computed and to be displayed
briefly. If $|q_n| < 1$, allow computation to continue by
doing nothing. The machine will stop by displaying a_n.
Press

$$STO-5 \quad \text{if } n = 4k+1$$
$$STO-4 \quad \text{if } n = 4k+2$$

$$STO+5 \quad \text{if } n = 4k+3$$
$$STO+4 \quad \text{if } n = 4k+4$$

Pressing

$$R/S$$

will cause the computer to proceed and to calculate next q_n and next a_n. If $|q_n| \overset{\geq}{=} 1$, quickly press

$$R/S$$
$$GTO\ 29$$
$$R/S$$

At next stop, the calculator will display asymptotic value of $J_\nu(x)$. To get asymptotic value of $Y_\nu(x)$, press

$$x \overset{>}{<} y$$

Caution: If $R < 0$ and $|S/R| < 10^{-5}$, the value of ϕ will be in error by precisely π, and signs of both J_ν and Y_ν will be wrong.

6. Examples and Timing

[1] $\nu = 0$, $x = 10$. Stopping at $n = 20$ yields

$$J_0(10) \approx -0.24593576 \ , \quad Y_0(10) \approx 0.05567117 \ .$$

Both values agree with the exact values to all digits. Total computing time about 95 sec.

[2] $\nu = \dfrac{1}{2}$, x = 2 :

$J_{1/2}(2) = 0.51301614$, $Y_{1/2}(2) = 0.23478571$.

These are the exact values.

[3] $\nu = \dfrac{11}{2}$, x = 5 :

$J_{11/2}(5) = 0.19056437$, $Y_{11/2}(5) = -0.57174942$.

These values agree with those computed by the program Bessel functions, arbitrary order.

RIEMANN ZETA FUNCTION ON CRITICAL LINE

1. Purpose

To compute the function

$$\xi(t) := \eta\left(\frac{1}{2} + it\right)$$

for real values of t, where

$$\eta(s) := \frac{s(s - 1)}{2} \; \Gamma\left(\frac{s}{2}\right) \; \pi^{-\frac{s}{2}} \; \zeta(s) \; ,$$

and ζ denotes the Riemann zeta function,

$$\zeta(s) := \sum_{n=1}^{\infty} n^{-s} \qquad (\operatorname{Re} s > 1) \; .$$

The function ξ is entire, even, and real for t real. There holds $\xi(t) = 0$ if and only if $s = \frac{1}{2} + it$ is a nontrivial zero of $\zeta(s)$ (see ACCA II, § 10.8).

2. Method

We use the integral representation

$$\xi(t) = \int_0^\infty f(x) \cos(2tx)\, dx \, , \qquad (1)$$

where $f(x) := \phi''(x) - \phi(x)$,

$$\phi(x) := e^x \, \omega(e^{4x}) \, ,$$

$$\omega(u) := \sum_{n=1}^\infty e^{-n^2 \pi u}$$

(see Bieberbach, Funktionentheorie Vol. II, Chapter 8).

The integral (1) is approximated by the midpoint formula, using an arbitrarily chosen step h, that is,

$$\xi(t) \doteq h \sum_{k=0}^{(\infty)} f(x_k) \cos(2tx_k) \, , \qquad (2)$$

where $x_k := \frac{1}{2}h + kh$. The summation is terminated as soon as $|f(x_k)| < \varepsilon$. The function f is evaluated by means of the series

$$f(x) = 8 \sum_{n=1}^{(\infty)} n^2 \pi \, e^{4x} (2n^2 \pi e^{4x} - 3) e^{x - n^2 \pi e^{4x}} \, . \qquad (3)$$

The summation is stopped when the general term (which is always positive) becomes $< \varepsilon$. The series converges rapidly for all $x \gtrless 0$.

Because of lack of programming space, no automatic step refinement is built into the program. However,

the step should always be refined manually to check
the accuracy of values. Because $f(x) = O(\exp ce^{-4x})$
as $x \to \infty$, the error of the numerical values of the
integral (1) is $O(h^m)$ for every $m > 0$, and Romberg
acceleration is therefore neither required nor effec-
tive.

3. Flow Diagram

We denote the sum in (2) by Σ, the sum in (3) by S,
and the n-th term of S by a_n.

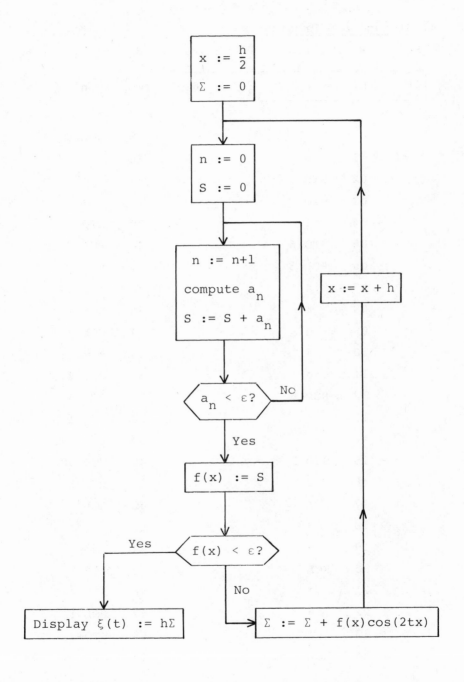

4. Storage and Program

R_0	R_1	R_2	R_3	R_4	R_5	R_6	R_7
2t	h	ε	$\dfrac{h}{2}$	0	temp	n	s
			x	Σ			

	00			25	RCL 5
→	01	0		26	−
	02	STO 6		27	e^x
	03	STO 7		28	*
→	04	1		29	STO+7
	05	STO+6		30	RCL 2
	06	RCL 3		31	x < y
	07	4		32	GTO 04
	08	*		33	RCL 2
	09	e^x		34	RCL 7
	10	π		35	$x \overset{>}{=} y$
	11	*		36	GTO 41
	12	RCL 6		37	RCL 4
	13	x^2		38	RCL 1
	14	*		39	*
	15	STO 5		40	GTO 00
	16	2	→	41	RCL 0
	17	*		42	RCL 3
	18	3		43	*
	19	−		44	cos
	20	RCL 5		45	*
	21	*		46	STO+4
	22	8		47	RCL 1
	23	*		48	STO+3
	24	RCL 3		49	GTO 01

5. Operating Instructions

Load the program, and switch to RUN. Press

SCI 8

to get the floating eight-digit display of results.
Load the data as follows:

$$2t \quad \text{into} \quad R_0$$
$$h \quad \text{into} \quad R_1$$
$$\varepsilon \quad \text{into} \quad R_2$$
$$\tfrac{1}{2}h \quad \text{into} \quad R_3 \ .$$

Clear R_4 by

CLX

STO 4

Because argument of cosine is in radians, press

RAD

When

PRGM

R/S

is pressed, the calculator will start whirling and

stop by displaying $\xi(t)$ as computed by formula (2). To check accuracy, the computation should be repeated with h $:= \frac{1}{2}$h. The necessary new data are furnished by

RCL 1	2
2	÷
÷	STO 3
STO 1	CLX
	STO 4

Note: Some of the foregoing instructions can be automated on a calculator with a few more memory locations.

6. Examples and Timing

Using $\varepsilon := 10^{-12}$, we get for

t = 14	h = 0.2	→ ξ =	-0.46487914	(24 sec)
	0.1		+0.00063699	(49 sec)
	0.05		0.00020129	(99 sec)
	0.025		0.00020129	(198 sec)

t = 15	h = 0.1	→ ξ =	-0.00001633
	0.05		-0.00070570
	0.025		-0.00070570

Thus a zero exists between t = 14 and t = 15, implying

the existence of a zero of the Riemann zeta function on the critical line between $s = \frac{1}{2} + 14i$ and $s = \frac{1}{2} + 15i$.

INDEX